AF324244

Beyond Metallocenes

ACS SYMPOSIUM SERIES **857**

Beyond Metallocenes

Next-Generation Polymerization Catalysts

Abhimanyu O. Patil, Editor
ExxonMobil Research and Engineering Company

Gregory G. Hlatky, Editor
Equistar Chemical LP

Sponsored by the
**ACS Division of Polymeric Materials:
Science and Engineering, Inc.**

American Chemical Society, Washington, DC

Library of Congress Cataloging-in-Publication Data

Beyond metallocenes : next-generation catalysts / Abhimanyu O. Patil, editor, Gregory G. Hlatky, editor ; sponsored by the ACS Division of Polymeric Materials: Science and Engineering, Inc..

    p. cm.—(ACS symposium series ; 857)

Includes bibliographical references and index.

ISBN 0–8412–3838–3

   1. Alkenes—Congresses. 2. Polyolefins—Congresses. 3. Polymerization—Congresses. 4. Catalysis—Congresses

  I. Patil, Abhimanyu O., 1952- II. Hlatky, Gregory G., 1956- III. American Chemical Society. Division of Polymeric Materials: Science and Engineering, Inc. IV. American Chemical Society. Meeting (223rd : 2002 : Orlando, Fla.) V. Series.

QD305.H7B49   2003
547′.28—dc21   2003050200

The paper used in this publication meets the minimum requirements of American National Standard for Information Sciences—Permanence of Paper for Printed Library Materials, ANSI Z39.48–1984.

PRINTED IN THE UNITED STATES OF AMERICA

Foreword

The ACS Symposium Series was first published in 1974 to provide a mechanism for publishing symposia quickly in book form. The purpose of the series is to publish timely, comprehensive books developed from ACS sponsored symposia based on current scientific research. Occasionally, books are developed from symposia sponsored by other organizations when the topic is of keen interest to the chemistry audience.

Before agreeing to publish a book, the proposed table of contents is reviewed for appropriate and comprehensive coverage and for interest to the audience. Some papers may be excluded to better focus the book; others may be added to provide comprehensiveness. When appropriate, overview or introductory chapters are added. Drafts of chapters are peer-reviewed prior to final acceptance or rejection, and manuscripts are prepared in camera-ready format.

As a rule, only original research papers and original review papers are included in the volumes. Verbatim reproductions of previously published papers are not accepted.

ACS Books Department

Contents

Catalysts Based on Copper Complexes

Indexes

Preface

There's no gainsaying that metallocene catalysts have had a stunning influence on the mature field of olefin polymerization. Compared to poorly understood heterogeneous conventional Ziegler–Natta systems that produce polymers of broad molecular weight distribution, metallocene catalysts exhibit a truly single-site nature, have extraordinarily high activities, and are amenable to rational modification and even to comprehensive characterization.

Metallocene polymerization catalysts were discovered almost at the same time as conventional Ziegler–Natta catalysts, but their low activities relegated them to playthings of the laboratory until they were joined with methylalumoxane to afford systems of much higher activity. The discovery that changes in substituents on the cyclopentadienyl ligand could influence activity, molecular weight, comonomer incorporation, and stereospecificity gave a further boost to their commercial applications.

But what is so special about the cyclopentadienyl group as a ligand for metallocene catalysts? It and the substituent groups on it serve to stabilize the active catalytic site and to exert a steric and electronic influence that affects catalyst performance and the characteristics of the polymer produced. By acting as stable, non-reactive ligands, the cyclopentadienyl groups prevent the active catalytic species from forming other active sites, leading to polymers with broad molecular weight and composition distributions.

An increasing recognition exists that these roles can more than adequately be filled by many other stable ancillary ligands. This and the already crowded, murky patent picture in metallocene catalysis have impelled investigators both in academia and industry to seek new single-site catalysts that share the advantages of metallocene systems, but that do not contain cyclopentadienyl groups. We term this new family of catalysts *next-generation* single-site olefin polymerization catalysts.

Broadly speaking, the stabilizing ancillary ligands used in these catalysts can be broken down into three major categories:

- isolobal equivalents to the cyclopentadienyl ligand, such as borata-benzenes, azaborolinyls and phospholes
- neutral ligands, such as the family of alpha-diimines
- anionic chelating ligands such as the phenoxyimines

The combined effect of these discoveries has been to broaden enormously our previous concepts of what can be done with olefin polymerization catalysts. Up to now, Group 4 metals (Ziegler–Natta or metallocene) produced linear high polymers and metals such as nickel were suitable only for oligomerization of olefins. Some next-generation single-site catalysts of the Group 4 metals already exhibit activities as high as their metallocene cousins and can produce living and stereo-regular polyolefins. Nickel-based next-generation catalysts polymerize ethylene to high molecular weight polymer with unusual hyper-branched structures. The portfolio of metals complexes capable of polymerizing ethylene has even been extended to copper. And next-generation cata-lysts are assisting in the assault of that most formidable of obstacles, the copolymerization of alpha-olefins with polar comonomers.

A symposium was organized by the Division of Polymeric Materials: Science and Engineering, Inc. at the 223rd American Chemical Society (ACS) National Meeting in Orlando, Florida on April 7–8, 2002 to review the scope of progress in this rapidly expanding field. The chapters in this volume are a result of presentations made at this symposium. To our knowledge, this is the first compilation of progress in next-generation non-metallocene polymerization catalysts. The reader will note the broad sweep in the metal complexes (from titanium to copper) used. New ligand synthesis, new metal compounds, new polymer struc-tures, and new combinations of monomers are all featured. In line with the *single-site* nature of these catalysts, chapters on the analysis and pre-diction of catalyst behavior using spectroscopy and molecular modeling are also included.

The editors thank the Division of Polymeric Materials: Science and Engineering, Inc., ExxonMobil Corporation, and Equistar Chemicals LP for their support of the symposium on which this book is based.

Ahhimanyu O. Patil
ExxonMobil Research and Engineering Company
1545 Route 22 East
Annandale, NJ 08801
Telephone: 908–730–2639
Fax: 908–730–2539
Email: abhimanyu.o.patil@exxonmobil.com

Gregory G. Hlatky
Equistar Chemicals LP
11530 Northlake Drive
Cincinnati, OH 45249
Telephone: 513–530–4004
Fax: 513–530–4206
Email: gregory.hlatky@equistarchem.com

Chapter 1

Beyond Metallocenes: Next-Generation Polymerization Catalysts

Abhimanyu O. Patil[1] and Gregory G. Hlatky[2]

[1]Corporate Strategic Research, ExxonMobil Research and Engineering Company, Route 22 East, Annandale, NJ 08801
[2]Equistar Chemicals LP, Cincinnati Technology Center, 11530 Northlake Drive, Cincinnati, OH 45249

This chapter surveys the recent development of new non-metallocene polymerization catalysts, late transition-metal polymerization catalysts and the polyolefins obtained using these catalysts. The effective homogeneous transition metal catalysts for olefin polymerization, especially those of Group 4 metals, must be one of the useful technologies for polymer synthesis in the two decades. These catalysts not only improved the property of manufactured polyolefins but also made it possible to synthesize new type of polyolefins. Over the past few years, trends in catalyst development have moved from modification of Group 4 metallocene catalysts to new generation catalysts, including non-metallocene catalysts and late transition-metal catalysts. Some of these new catalysts can produce novel polyolefin microstructures and functional polyolefins, which are not possible with metallocene catalysts.

Polyolefins are by far the largest class of synthetic polymers made and used today. There are several reasons for this, such as low cost of production, lightweight, and high chemical resistance. Most of these polyolefins are manufactured using Ziegler-Natta catalysts, chromium-based catalysts and the free-radical process for low-density polyethylene (LDPE) (*1-4*) In the free-radical polymerization of ethylene, short- as well as long-chain branches are formed without any added co-monomer. Short-chain branches affect polymer properties, such as crystallinity and melting temperature, are important in controlling polyolefin application properties. Long-chain branching influence the rheology of polyolefin melts, and result in good processing properties of LDPE (*4*). Other attractive feature of the high-pressure free radical process is the possibility of incorporating functionalized olefins, such as vinyl acetate or acrylates.

Significant portions of these polyolefins are manufactured with Ziegler-Natta catalysts. These commercial heterogeneous catalysts exhibit very high activity for ethylene and propylene polymerization. What seems to be lacking, however, is homogenization of active species and precise control of polymerization. Kaminsky and Sinn developed the catalyst system which was composed of group 4 metallocene catalysts combined with MAO and afforded to give high activity for ethylene polymerization (*5*). The active species in the catalyst system were homogeneous and produced polyethylene with narrow molecular weight distribution. Since the discovery of Cp_2ZrCl_2-MAO catalyst system for ethylene polymerization, much effort has been made to the improvement of the metallocene catalyst (*6*) Trend in catalyst development for the polymerization is moving from modification of metallocene or mono-cyclopentadienyl type catalysts to designing of non-metallocene catalysts, in short, the catalysts without cyclopentadienyl ligands.

One of the objectives of research on olefin polymerization by late transition metal complexes is the design of catalysts capable of producing high molecular weight products. The polymerization activity with the non-metallocene catalysts is generally lower than that with the metallocene catalyst. However, some non-metallocene catalysts, which show high activity for ethylene polymerization, have been developed. For example, Mitsui Chemicals developed zirconium and titanium complexes (**1**) involving a bis(salicylaldiimine) ligand (*7-8*) The polymerization activity with a zirconocene catalyst is twenty times as high as that with a Cp_2ZrCl_2 catalyst. Gibson *et al.* (*9-10*) and Brookhart *et al.*(*11*) reported ethylene polymerization with Fe(II) complexes (**2**). The complexes are based on a five-coordinate iron center supported by a neutral tridentate 2,6-bis(imino)pyridine ligand. The catalysts, on activation with methylaluminoxane (MAO) showed high ethylene polymerization activity.

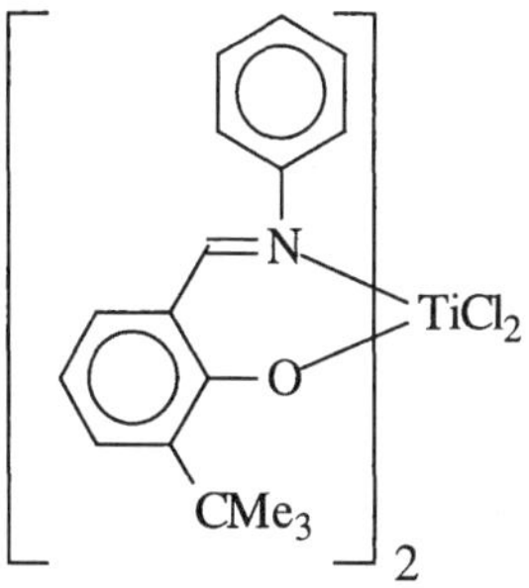

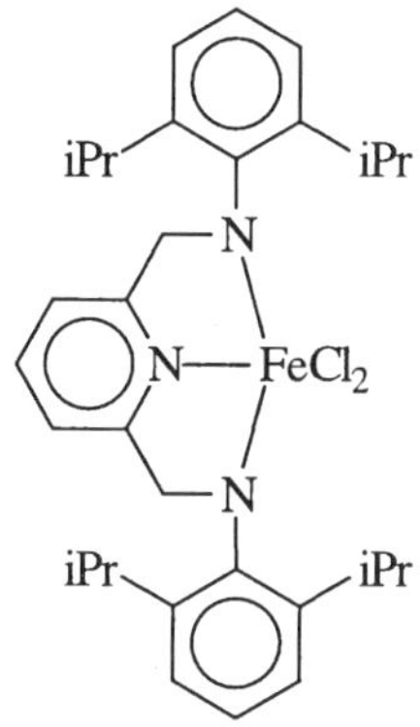

1

2

Scheme 1. Metal complexes for ethylene polymerization with bis(salicylaldimine) ligands (1) and 2,6-bis (imino) pyridine ligands (2).

The synthesis of linear copolymers of olefins and polar monomers such as acrylates and vinyl acetate has been an elusive goal for polymerization chemists (*12*) Polar monomers, such as acrylates or vinyl acetate tend to poison Ziegler-Natta and metallocene catalysts. Currently such olefin polar copolymers are produced by free radical polymerization procedures that tend to provide somewhat branchy copolymers. There are great advances have been made in recent years in the filed of mid- and late-transition metal polymerizations.

The catalysts reported in the literature to this point are based on Group 8-11 metals (Fe, Co, Ni, Pd, Cu) (*13-17*). The lower oxophilicity and presumed greater polar group tolerance of late transition metals relative to early metals make them attractive catalysts for the copolymerization of ethylene with polar comonomers under mild polymerization conditions. Late transition metal catalysts (*18, 19*), in general, are less sensitive to polar species. However, the known late metal catalysts still have difficulty preparing linear functional polyolefins.

Pd(II) and Ni(II) Catalyst Complexes with Multidentate Ligands

In mid nineties Brookhart and co-workers reported on the discovery of a new class of catalysts for the polymerization of ethylene, α-olefins, cyclic olefins and the copolymerization of nonpolar olefins with polar monomers (*20-27*). These catalysts, based on known nickel(II) and palladium(II) complexes of bulky substituted diimine ligands, are unique in polymerizing ethylene to highly

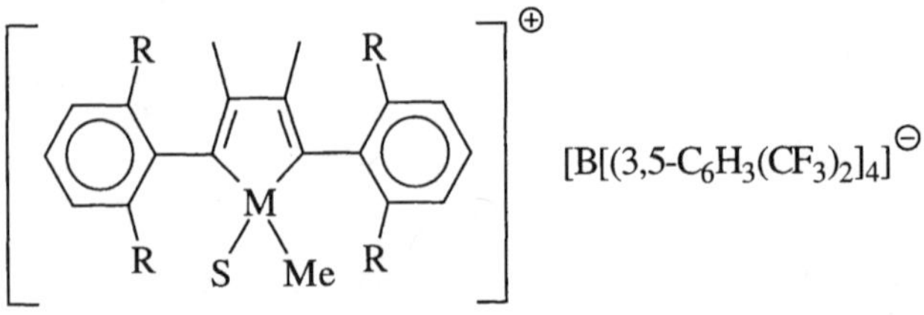

$$[\text{B}[(3,5\text{-}C_6H_3(CF_3)_2]_4]^{\ominus}$$

3

Scheme 2: Structure of the active site of a late-transition-metal diimine complex 3. S = solventing molecule, M = Ni, Pd.

branched, high molecular weight homopolymers at remarkable reaction rates. Activities of up to 4×10^6 TO h^{-1} were reported for the cationic Ni(II) catalysts, which thus approach rates typical of metallocenes (*28-29*) The key features of the α-diimine polymerization catalysts are (1) highly electrophilic, cationic nickel and palladium metal centers; (2) the use of sterically bulky α-diimine ligands.

Polymer molecular weight distributions are relatively narrow and lie in the range expected for single-site catalysts. A key feature enabling polymerization to occur to very high molecular weights is retardation of chain transfer by the steric bulk of the *o*-aryl substituents (R = iPr or Me). The R groups are located in an axial position above and below the square-planar coordination plane as a consequence of a perpendicular arrangement of the aryl rings with respect to the latter. Introduction of less bulky ortho substituents (R=H) in the Ni(II) catalyts yields a highly active cationic catalyst for oligomerization of ethylene to linear α-olefins.

The polyethylenes possess even- and also odd-number carbon branches (Me, Et, Pr, Bu, amyl, and longer branches). Whereas the polymers obtained with Ni(II) catalyts are dominated by methyl branches, the Pd(II) catalysts yield a large portion of longer branches and even a branch on a branch moiety, that is, an element of a hyperbranched structure. The Pd(II) catalysts yield highly branched amorphous polymers (approximately 100 branches per 1000 carbon atoms) irrespective of the ethylene concentration, whereas branching can be controlled with the Ni(II) catalysts by varying the ethylene concentration and polymerization temperature (1 to 100 branches per 1000 carbon atoms). Accordingly, the macroscopic properties of these polyethylenes - obtained from ethylene as the sole monomer - range from highly viscous (high molecular weight) liquids with a T_g of -70 °C, over rubbery elastomeric materials, to rigid linear polyethylenes.

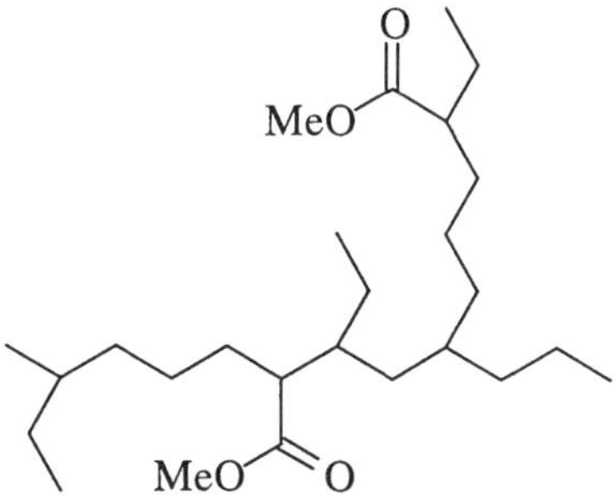

Scheme 3. Representative structure of a highly branched random copolymer of ethylene and methyl acrylate prepared by palladium catalysts.

Remarkably, the Pd(II) catalysts also allow for the coordination copolymerization of ethylene and simple polar functionalized olefins, such as acrylates (Scheme 3). As the amount of acrylate is increased from 1 to 12 mole%, the catalyst turnover numbers decrease from 7710 mole of ethylene/mole of catalysts to 455 and With increasing acrylate incorporation into the copolymer, catalyst activity is lowered as a consequence of a reversible blocking of coordination sites by chelating binding of the carbonyl groups of incorporated acrylate. Chain running also results in predominant incorporation of acrylate at the ends of branches in the highly branched, amorphous copolymers. These catalyst system do not incorporate conventional polar monomers such as acrylates into the backbone of a linear polyethylene.

Tridentate Iron/Bis(imino)pyridyl Catalyst Complexes

One of the major developments in the area of late-transition metal polymerization catalysis is the use of novel iron- and cobalt-based catalysts reported by Brookhart and Gibson (*9-11*). These five-coordinate precursor complexes of tridentate iron (II) with 2,6-bis(imino)pyridyl ligands are reported to be very active for the oligmerization and polymerization of ethylene. The ligands were prepared by the Schiff-base condensation of two equivalents of substituted aniline with 2,6-diacetylpyridine. Steric effects and the electronic factor of ligands affect the catalytic activity, behavior of catalysts as ethylene oligomerization and/or polymerization catalysts. The bulk around the metal center is a key in retarding chain transfer to obtain a high molecular weight polymer.

Astounding rates of up to 10^7 TO h^{-1} (at 40 bar ethylene pressure) have been reported for the iron catalysts. The typical relative ease of preparation for many of the late transition metal catalyst systems highlighted is exemplified by the

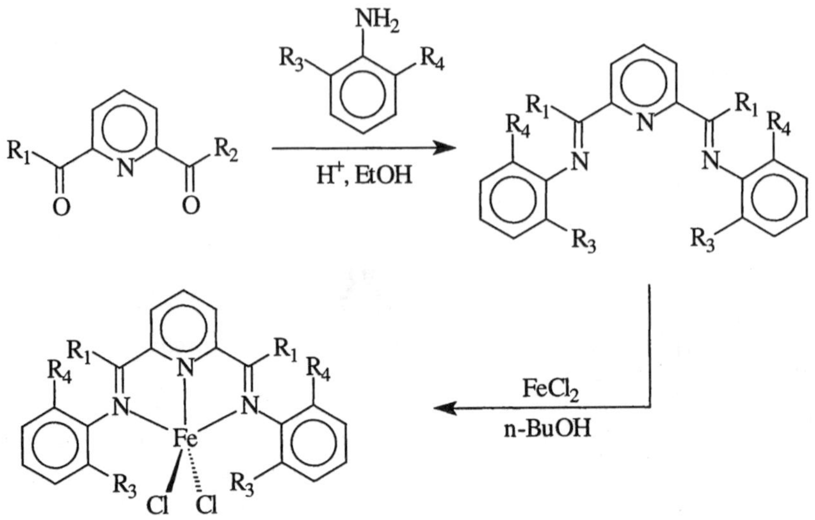

Scheme 4. Synthesis of iron bis(imino)pyridyl complexes.

preparation of the iron catalysts from commercially available compounds as shown in Scheme 4. Broad or bimodal molecular weight distributions are obtained with the Fe catalysts as a result of chain transfer to aluminum as well as β-H elimination. In analogy to the previously mentioned diimine-substituted Ni(II) and Pd(II) catalysts, the steric bulk of the *o*-aryl substituents retards chain transfer. Reducing the steric bulk of the *o*-aryl substituents R in the iron complexes again yields catalysts for ethylene oligomerization (*30-31*). In contrast to the nickel and palladium systems, there is no chain walking and the polyethylene is strictly linear with very high density. Astonishing activities of up to 10^8 TO h^{-1} for formation of >99 % linear α-olefins were observed, far exceeding the activities reported for SHOP catalysts.

Neutral Nickel(II) Catalysts with N^O Chelating Ligands

The copolymerization of ethylene with a variety of functional monomers using Ni complexes containing phosphorus-oxygen chelating ligands were reported by Klabunde and Ittel (*32*). Even though vinyl acetate and methyl methacrylate did not undergo compolymerization, but copolymers could be obtained if the functional group and vinyl double bond were separated by at least two methylene spacers in the polar monomer.

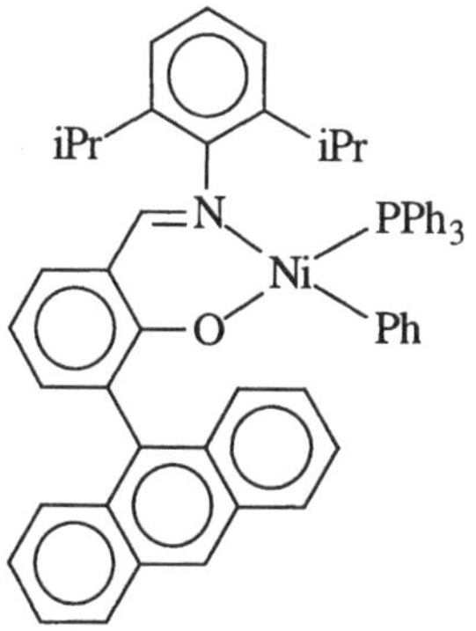

Scheme 5: Structure of a salicylaldimido nickel polymerization catalyst.

A new class of neutral nickel catalysts based on salicylaldimine ligands (*33*) was reported independently by Johnson *et al.* (*34*) and Grubbs and co-workers (*34-35*). These neutral nickel(II) complexes (Scheme 5) contain a formally monoanionic bidentate ligand and are thus related to SHOP-type catalysts. However, the P donor of the latter is replaced by a bulky substituted imine moiety. Moderately branched to linear polyethylenes are obtained with such catalyst systems. Similar weight-average molecular weights but considerably narrower molecular weight distributions are obtained than in materials prepared with traditional P,O-based catalysts. Initially, substituted allyl complexes (*34*) or triphenylphosphane complexes (*35*) were used, which require a Lewis acid or phosphane scavenger, respectively, as a cocatalyst for effective polymerization. Introduction of bulky substituents R' in the *ortho* position of the phenolate moiety enables the use of phosphane complexes as single-component precursors without a phosphane scavenger as a result of facilitated dissociation of the triphenylphosphane ligand (*36*).

Activities in ethylene homopolymerization (2.3×10^5 TO h^{-1} at 17 bar) equal those achieved with optimized P,O-based polymerization catalysts. Like P,O-based neutral nickel(II) catalysts, the N,O-substituted nickel systems are functional-group tolerant. Grubbs and co-workers have copolymerized ethylene with 5-functionalized norbornenes (5-norbornen-2-yl acetate or 5-norbornen-2-ol) (*36*), and high molecular weight polymer was obtained in ethylene homopolymerization in the presence of added polar reagents, such as ethers, ethyl acetate, acetone, and also water (*36*). The maximum comonomer incorporation was 22 and 12 wt% for NB-OH and NB-COOH, respectively (Scheme 6). The tolerance of polar groups was tested by running ethylene polymerization in the presence of small molecule additives at a level of ~ 1500 equivalent per metal center. Both decreased in molecular weight and catalysts turn over numbers were observed with different polar additives. Such

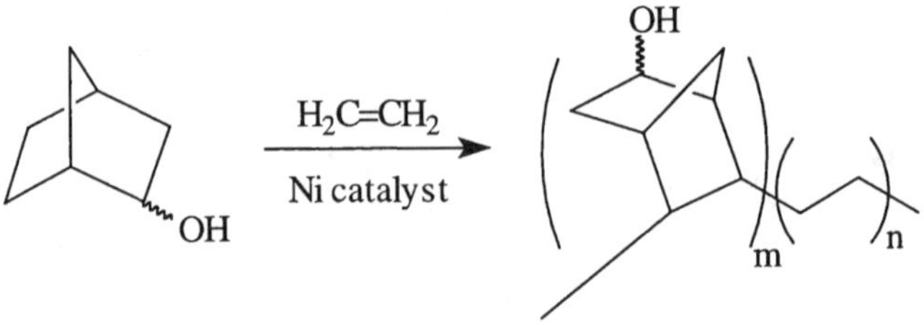

Scheme 6. Copolymerization of ethylene with functional norbornene using Ni catalyst.

polymerizations in the presence of added polar reagents are of interest with regard to polymerization of less rigorously purified monomers (*36*).

Cu Bis-benzimidazole Catalysts

Olefin polymerization catalysts based on the more electron-rich Group 11 metals have been recently prepared (*37-41*). The Cu bis-benzimidazole pre-catalysts (activated by MAO) not only homopolymerize both ethylene and various acrylates but also copolymerize these two monomer classes. The ability of these catalysts to randomly copolymerize olefins with polar monomers via a metal-mediated mechanism was unprecedented.

Homopolymers of ethylene prepared by Cu BBIM/MAO at 700-800 psig ethylene, 80°C and Al/Cu ratio of > 200/1 were typically high molecular weight and high melting polymers (Tm > 135 °C). The ^{13}C NMR of HDPE prepared by Cu BBIM/MAO contains no detectable branches. This is in contrast to the first generation Brookhart catalysts that often produce polymers with 80-150 branches / 1000 carbons. Homopolymerization of acrylates by Cu BBIM/MAO was typically run at lower temperatures (~ 25°C) and ratios of Al/Cu (< 50). The order of reactivity was *t*-BuA > *n*-BuA > MMA > E. CuBBBIM/MAO does not homopolymerize traditional free radical polymerizable monomers like styrene, vinyl acetate or butadiene. Also, unlike Brookhart bis-imine catalysts, it does not homopolymerize cyclopentene.

Ethylene and *t*-butyl acrylate have been copolymerized at 600 - 800 psig and 80°C using Cu BBIM/MAO catalysts (Scheme 7). In contrast to the first generation Brookhart di-imine Pd catalysts, high levels of acrylate incorporation was possible. Also, high molecular weight, narrow MWD copolymers with low level of branching can be made. The products were shown to be true copolymers by ^{13}C NMR and GPC/UV. The presence of EAE, and EAA/AAE triads in the Cu BBIM/MAO catalyzed E/*t*-BuA copolymers was a clear indication of copolymer formation. Additional support for the presence of copolymers was a

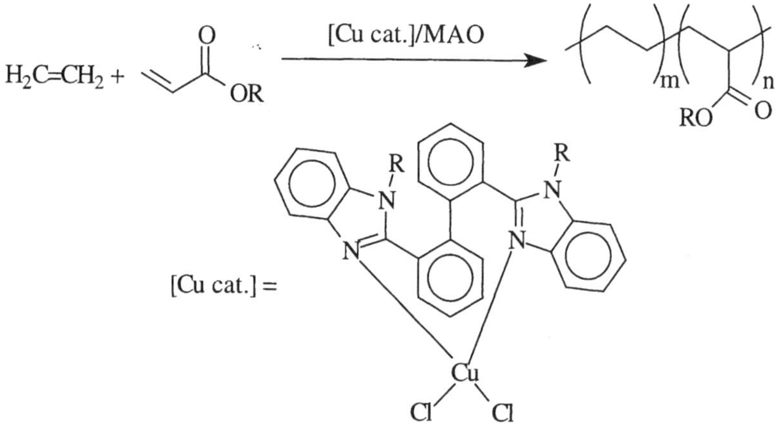

Scheme 7. Cu BBIM/MAO catalyzed ethylene/acrylate copolymers.

uniform distribution of UV activities across the MWD in the GPC/UV trace. Unlike Brookhart catalyzed ethylene/acrylate copolymers, the Cu catalyzed copolymers are in-chain copolymers with high levels of acrylate incorporation. Thus, these cu catalysts represents the first transition metal system capable of carrying out true random copolymerizations of polar vinyl monomers with non-polar olefins. Low to high molecular weight products were prepared.

While the mechanism of Cu BBIM/MAO polymerizations is not fully determined, the ethylene homopolymerization data were consistent with a single-site coordination/insertion mechanism based upon the high degree of linearity and narrow MWD in the product. The acrylate homopolymerization and copolymerization cases may proceed by a different or additional mechanism. However, conventional free-radical or ATRP mechanisms are less likely. Conventional free radical chemistry was ruled out because Cu BBIM/MAO did not polymerize other traditional free radical monomers (e.g., styrene, butadiene, vinyl acetate). Moreover, Cu BBIM/MAO polymerizes monomers in the presence of inhibitors. Recently, copper α-diimine complexes containing 2,6-diphenylimino substituents were found to polymerize ethylene using methylaluminoxane cocatalyst (*42*).

References

1. Ziegler Catalysts, Eds: G. Fink, R. Mulhaupt, H. H. Brintzinger, Springer, Berlin 1995.
2. Brintzinger, H. H.; Fischer, D.; Mulhaupt, R.; Rieger, B.; Waymouth, R. *Angew. Chem. Int. Ed. Engl.* **1995**, *34*, 1143.

3. Kaminsky, W.; Arndt, M. Adv. Polym. Sci. **1997**, *127*, 143.

4. Whiteley, K. S. in *Ullmann's Encyclopedia of Industrial Chemistry*, Vol. A21, 5th ed. Eds: Gerhartz, W.; Elvers, B., VCH, Weinheim, pp. 488-517, 1992.]

5 Sinn, H.; Kaminsky, W.; Vollmer, H. J.; Woldt, R. *Angew. Chem. Int. Ed. Engl.* **1980**, *19*, 369.

6 Imanishi, Y.; Naga, N. *Prog. Polym. Sci.* **2001**, *26*, 1147.

7 Matusi, S.; Tohi, Y.; Mitani, M.; Saito, J.; Makio, H.; Tanaka, H.; Nitabaru, M.; Nakano, T.; Fugita, T. *Chem. Lett.* **1999**, 1065.

8. Matusi, S.; Mitani, M.; Saito, J.; Tohi, Y.; Makio, H.; Tanaka, H.; Fugita, T. *Chem. Lett.* **1999**, 1263.

9 Britovsek, G. J. P.; Gibson, .V. C.; Kimberley, B. S.; Maddox, P. J. McTavish, S. J.; Solan, G. A.; White, A. J. P.; Williams, D. J. *Chem. Commun.* **1998**, 849-850.

10. Britovsek, G. J. P.; Bruce, M. I.; Gibson, V. C.; Kimberley, B. S.; Maddox, P. J.; Mastroianni, S.; McTavish, S. J.; Redshaw, C.; Solan, G. A.; Strömberg, S.; White, A. J. P.; Williams, D. J. *J. Am. Chem. Soc.* **1999**, *121*, 8728-8740.

11. Small, B. L.; Brookhart, M.; Bennett, A. M. A. *J. Am. Chem. Soc.* **1998**, *120*, 4044050.

12. D. N. Schulz, A. O. Patil in "Functional Polymers - Modern Synthetic Methods and Novel Structures", ACS Symposium Series 704, 1, 1998.

13. Ittel, S. D.; Johnson, L. K.; Brookhart, M. *Chem. Rev.* **2000**, *100*, 1169-1203.

14. Britovsek, G. J. P.; Gibson, V. C.; Wass, D. F. *Angew. Chem. Int. Ed.* **1999**, *38*, 428-447.

15. Boffa, L. S.; Novak, B. M. *Chem. Rev.* **2000**, *100*, 1479-1493.

16. Yanjarappa, M. J.; Sivaram, S. *Prog. Polym. Sci.* **2002**, *27*, 1347.

17. Mecking, S. *Angew. Chem. Int. Ed.* **2001**, *40*, 534.

18. Abu-Surrah, A.; Rieger, B. *Angew. Chem. Int. Ed.* **1996**, *35*, 2475.

19. Brookhart, M.; Wagner, M. I. *J. Am. Chem. Soc.* **1994**, *116*, 3641.

20. Johnson, L. K.; Killian, C. M.; Brookhart, M. *J. Am. Chem. Soc.* **1995**, *117*, 6414-6415.

21. Johnson, L. K.; Mecking, S.; Brookhart, M. *J. Am. Chem. Soc.* **1996**, *118*, 267-268.

22. Killian, C. M.; Tempel, D. J.; Johnson, L. K.; Brookhart, M. *J. Am. Chem. Soc.* **1996**, *118*, 11664-11665.

23. Killian, C. M.; Johnson, L. K.; Brookhart, M. *Organometallics* **1997**, *16*, 2005-2007.

24. Mecking, S.; Johnson, L. K.; Wang, L.; Brookhart, M. *J. Am. Chem. Soc.* **1998**, *120*, 888-899.

25. Svejda, S. A.; Johnson, L. K.; Brookhart, M. *J. Am. Chem. Soc.* **1999**, *121*, 10634-10635.

26. Gates, D. P.; Svejda, S. A.; Onate, E.; Killian, C. M.; Johnson, L. K.; White, P. S.; Brookhart, M. *Macromolecules* **2000**, *33*, 2320-2334.

27. Guan, Z.; Cotts, P. M.; McCord, E. F.; McLain, S. J. *Science* **1999**, *283*, 2059-2062.

28. Kaminsky, W.; Arndt, M. *Adv. Polym. Sci.* **1997**, *127*, 143-187

29. Alt, H. G.; Köppl, A. *Chem. Rev.* **2000**, *100*, 1205-1222.

30. Small, B. L.; Brookhart, M. *J. Am. Chem. Soc.* **1998**, *120*, 7143-7144.

31. Britovsek, G. J. P.; Mastroianni, S.; Solan, G. A.; Baugh, S. P. D.; Redshaw, C.; Gibson, V. C.; White, A. J. P.; Williams, D. J.; Elsegood, M. R. J. *Chem. Eur. J.* **2000**, *6*, 2221-2231

32. Klabunde, U.; Ittel , S. D. *J. Mol. Catal.* **1987**, *41*, 123.

33. Desjardins, S. Y.; Cavell, K. J.; Hoare, J. L.; Skelton, B. W.; Sobolev, A. N.; White, A. H.; Keim, W. *J. Organomet. Chem.* **1997**, *544*, 163-174. These authors have originally reported the ethylene polymerization using other N,O-substituted pyridine-carboxylato nickelII complexes. Howevere, the polymerization rate was comparatively low.

34. Johnson, L. K.; Bennett, A. M. A.; Ittel, S. D.; Wang, L.; Parthasarathy, A.; Hauptman, E.; Simpson, R. D.; Feldman, J.; Coughlin, E. B. PCT Int. Appl. WO 98/30609, 1998.

35. Wang, C.; Friedrich, S.; Younkin, T. R.; Li, R. T.; Grubbs, R. H.; Bansleben, D. A.; Day, M. W. *Organometallics* **1998**, *17*, 3149-3151.

36. Younkin, T. R.; Connor, E. F.; Henderson, J. I.; Friedrich, S. K.; Grubbs, R. H.; Bansleben, D. A. *Science* **2000**, *287*, 460-2.

37. Stibrany, R. T.; Patil, A. O.; Zushma. *Polym. Mater Sci. Eng.* **2002**, *86*, 323.

38. Stibrany, R. T.; Schulz, D. N.; Kacker, S.; Patil, A. O.; Baugh, L. S.; Rucker, S. P.; Zushma, S.; Berluche, E.; Sissano, J. A. *Polym. Mater. Sci. Eng.* **2002**, *86*, 325.

39. Stibrany, R. T.; Schulz, D. N.; Kacker, S.; Patil, A. O. U.S. Patent 6,037,297, 2000.

40. Stibrany, R. T.; Schulz, D. N.; Kacker, S.; Patil, A. O. U.S. Patent 6,417,303, 2002.

41. Stibrany, R. T., U.S. Patent 6,180,788, 2001.

42. Gibson, V. C.; Tomov, A.; Wass, D. F.; White, A. J. P.; Williams, D. J. *J. Chem. Soc. Dal. Trans.* **2002**, *11*, 2261-2262.

Catalysts from Group 4 Metal Complexes

Chapter 2

Syntheses of New Olefin Polymerization Catalysts Based on Zirconium Complexes of Organoboron Compounds

Arthur J. Ashe, III, Hong Yang, Xinggao Fang, Xiangdong Fang, and Jeff W. Kampf

Department of Chemistry, The University of Michigan, Ann Arbor, MI 48109–1055

New single-site olefin-polymerization catalysts have been prepared by the substitution of anionic organoboron heterocycles for Cp in Cp_2ZrCl_2. The syntheses of the diheterocyclopentadienyl ligands 1,2-thiaborolyl, 1,3-thiaborolyl and 1,2-azaborolyl are described. These ligands can serve as surrogates for Cp in Zr(IV) complexes, which on activation by methylaluminoxane are good catalysts for the polymerization of ethylene. Similiarly, the synthesis of the indenyl surrogate 3a,7a-azaborindenyl is described. 3a, 7a-Azaborindenyl forms a Zr(IV) complex which closely resembles the corresponding indenyl complex.

In the past decade investigation of group 4 metallocenes (**1**) has been one of the most active areas of organometallic research (*1*). The major interest is the conversion of these metallocenes into highly active single-site catalysts for the polymerization of olefins (*2*).These polymerizations occur at a single well-defined metal center, which makes it possible to correlate metallocene structure with polymer properties. Intense research efforts have focused on the design and synthesis of sterically defined metallocenes which in the most favorable cases are both highly active and stereoselective catalysts. The best known example is the C_2-symmetric bridged bis indenyl zirconocenes, e.g. **2**, originally developed by Brintzinger (*3*). The C_2-site of **2** induces chirality transfer to the polymer such that propylene is converted to isotactic polypropylene. This subject has been extensively reviewed (*2*).

In contrast, much less research has been done on modifications of the catalytically active metal site by the substitution of heterocyclic ligands in place of Cp. In 1995 we prepared the first boratabenzene zirconium derivatives **3**, **4** and **5** (*4,5*) and in collaboration with Professor G. Bazan established that methylaluminoxane (MAO) activated **3** is an excellent catalyst for polymerization of ethylene. This is a very important result since it shows that the π-bound boron heterocycle is tolerant to the aggressive Lewis acidic conditions of Ziegler-Natta polymerization and it can replace Cp in zirconium complexes without undiminishing the catalytic acitivity.

It was also found that MAO activated **5** reacted with ethylene to give only low MW oligomers. This result demonstrates that a change of the B-substituent can change reactivity centered at Zr. This difference appears to be a consequence of more rapid β-hydrogen elimination from the more electrophilic catalyst derived from **5**. Since prior work had shown that the B atom of boratabenzenes interacts strongly with exocyclic donor substituents (*6*) and since the B-substituents are easily changed, it appears possible to electronically tune boratabenzene complexes in a manner not possible for Cp-complexes (*4,5*).

Heteroborolyls

Current work is focusing on Zr and Ti complexes of an extended family of boron containing heterocycles, particularly thiaborolyls **6** and **7** and azaborolyls **8** and **9**. (See Scheme 1)

The formal replacement of a CH group by the isoelectric BH⁻ group converts a neutral aromatic to an anion. In this manner benzene becomes boratabenzene while thiophene is converted to **6** and **7** and pyrrole to **8** and **9** respectively.

Since thiophene and pyrrole are more π-electron rich than benzene, **6-9** are expected to be more electron rich than boratabenzene and metal complexes of **6-9** should be more electron rich than the corresponding boratabenzene complexes.

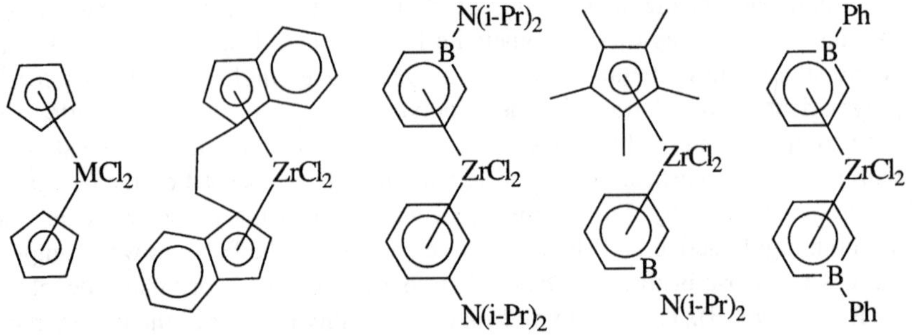

MCl₂   ZrCl₂   ZrCl₂   ZrCl₂   ZrCl₂

1   **2**   **3**   **4**   **5**

M = Ti, Zr, Hf

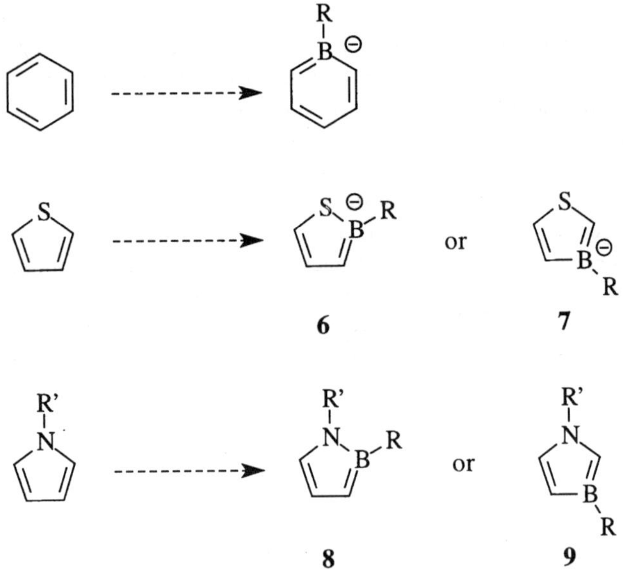

6   **7**

8   **9**

Scheme 1: Replacement of CH by BR⁻ converts an aromatic compound to an anion

Indeed it can be argued that exchanging a Cp ring for one of these heterocyclic rings will give complexes with different electronic properties which might complement those of the corresponding metallocenes.

A series of N-tert-butyl and N-trimethylsilyl-1,2-azaborolides **8a** were prepared by Schmid and coworkers who used **8** as a replacement ligand for Cp (See Scheme 2) (*7*).A report in the patent literature that Zr complex **10** is a good precatalyst for olefin polymerization is highly encouraging (*8*). However, the heterocycles **6**, **7** and **9** have not been reported in the prior literature. Therefore the major challenge in exploring this chemistry has been the development of syntheses of the ligands.

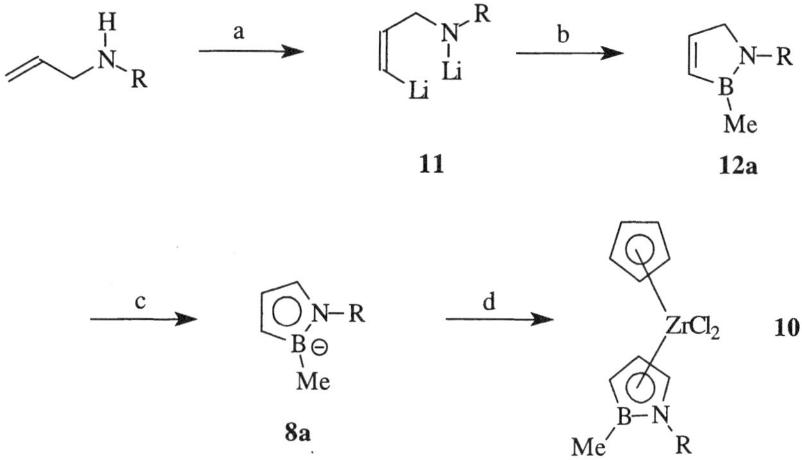

Scheme 2: The Schmid synthesis of 1,2-azaborolyllithium.

Key: a) BuLi/TMEDA, b) MeBBr$_2$, c) LTMP, d)CpZrCl$_3$

We have prepared N,N-diisopropyl-3-amino-1,3-thiaborolyllithium **7a** via a multistep synthesis, illustrated in Scheme 3 (*9*). The key step involved the ring closure of a 1,4-lilithio compound on a boron dihalide (**15**➡**16**). The subsequent reaction of **16** with t-butyllithium afforded **7a** which is easily converted to transition metal complexes. The reaction of **7a** with [Cp*RuCl]$_4$ gave Cp*Ru-complex **17**, the crystal structure of which shows it to be a diheteroruthenocene. The reaction of **7a** with Cp*ZrCl$_3$ afforded the Cp*ZrCl$_2$-complex **18**, which presumably has a bent-metallocene type of structure.

On activation by excess methylaluminoxane, **18** was active towards the polymerization of ethylene. Under identical conditions, the reactivity of **18** and **3** were found to 7.5 × 10^4 and 20.4 × 10^4 (g polymer/mol Zr·atm), respectively (*9*).The activity of **18** is approximately the same as the corresponding complex

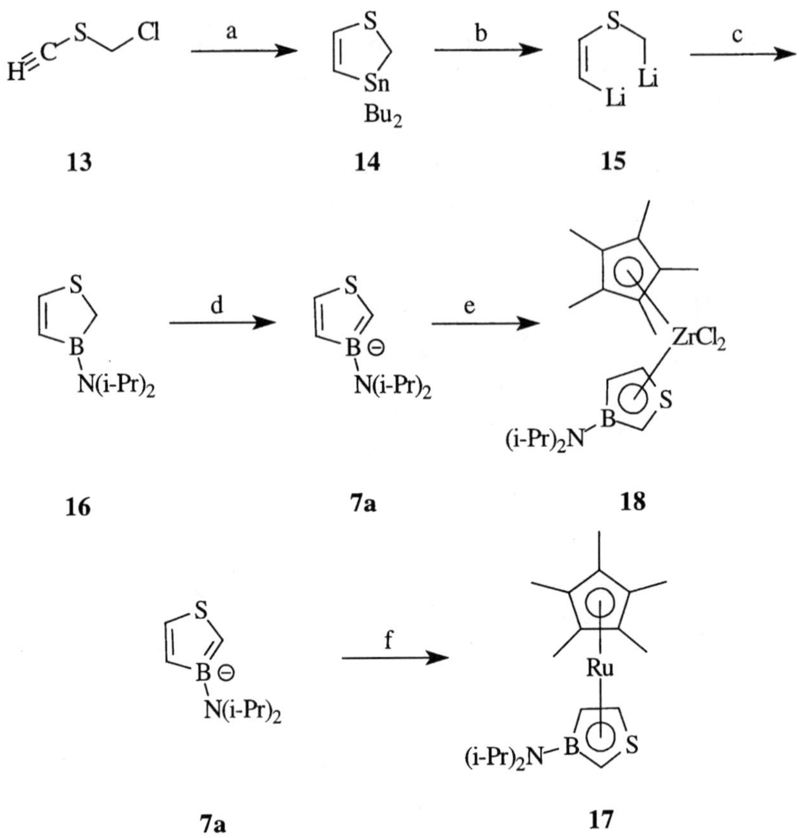

13   **14**   **15**

16   **7a**   **18**

7a   **17**

Scheme 3: Synthesis of 1,3-thiaborolyl

*Key: a) Bu$_2$SnH$_2$/LDA, b) BuLi, c) I-Pr$_2$NBCl$_2$, d) t-BuLi, e) Cp*ZrCl$_3$*

*f) (Cp*RuCl)$_4$*

of boratabenzene **4**. Thus **7** is a suitable replacment ligand for Cp in metallocene based olefin polymerization catalysts.

For comparison we wished to examine complexes of the isomeric 1,2-thiaborolyl **6**. However a projected synthesis based on a 1,4-dilithio compound failed in our hands (*10*).Fortunately an alternative synthesis based on the Grubbs Ring Closing Metathesis (RCM) was very successful. (See Scheme 4) (*11*)

(Allylthio)vinyl borane **20** is readily available from the reaction of allylmercaptan with N,N-diisopropylaminovinylboron chloride. Upon treatment of **20** with 1 mol % Grubbs catalyst $(Cy_3P)_2(PhCH)RuCl_2$ in CH_2Cl_2 at 25°C cyclization occurred in 95% yield to afford the heterocycle **21**. Deprotonation of **21** by LDA in ether gave **6a** in high yield.

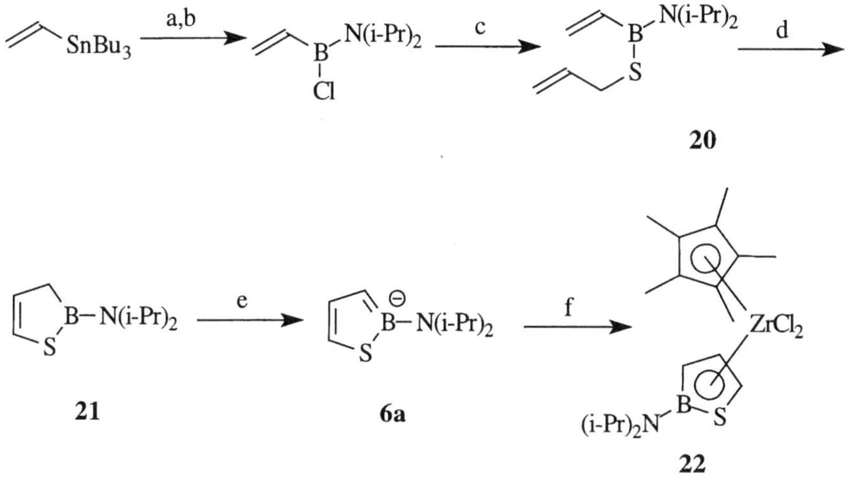

Scheme 4: Synthesis of 1,2-thiaborolyl using RCM

Key: a) BCl_3, b) $HN(i-Pr)_2$, c) C_3H_5SH/NEt_3, d) $(Cy_3P)_2(PhCH)RuCl_2$, e) LDA, f) $Cp*ZrCl_3$

Like its 1,3-isomer 1,2-thiaborolillithium readily forms transition metal complexes. Reaction of **6a** with $Cp*ZrCl_3$ gave the corresponding $Cp*ZrCl_2$-complex **22**. On activation by excess MAO **22** was an active catalyst towards polymerization of ethylene. The activity of **22** was only 20% less than that of **18**. Apparently the regiochemical difference between **18** and **22** is not of major consequence. It might also be noted that neither **18** nor **22** were very active catalysts for polymerization of α-olefins.

Since the RCM route to **6** was so efficient, it seemed desirable to apply it to the preparation of **8**. (See Scheme 5) (*12,13*). The necessary starting material N-Allyl, B-vinyl aminoboranes, e.g. **23**, are easily prepared by the reaction of secondary allylamines with substituted vinylboron halides. Upon treatment of **23** with 5 mol% Grubbs catalyst in CH_2Cl_2 cyclization occurred smoothly to afford

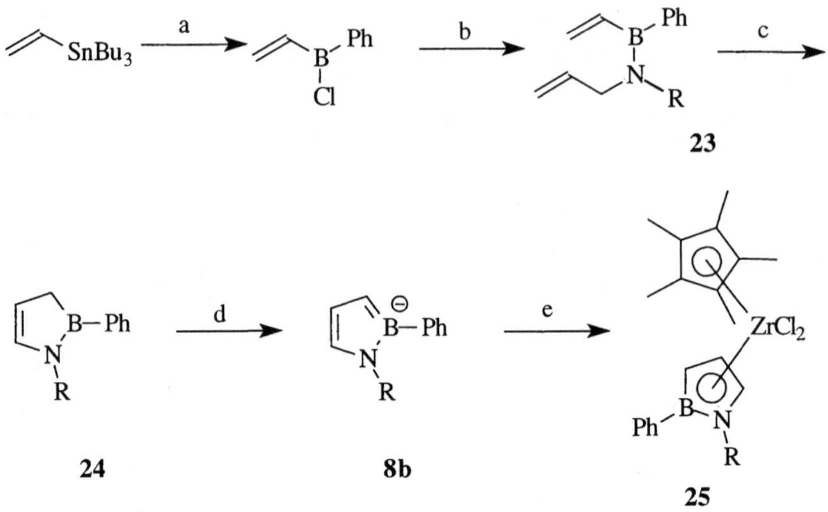

Scheme 5: Synthesis of 1,2-azaborolyl using RCM

Scheme 5: Synthesis of 1,2-azaborolyl using RCM

*Key: a) PhBCl$_2$, b) C$_3$H$_5$NHR/NEt$_3$, c) (PCy$_3$)$_2$(PhCH)RuCl$_2$, d) LDA, e) Cp*ZrCl$_3$*

24 in 80% yield. Subsequent reaction with LDA followed by Cp*ZrCl$_3$ afforded crystalline complex **25**. The molecular structure of **25**, illustrated in Figure 1, shows it to be a bent dihetero-metallocene.

Since the B- and N- substituents are easily changed, the synthesis outlined in Scheme 5 is more general than the Schmid synthesis. To illustrate this generality we have recently prepared a B,N-analog of indenyl 3a,7a-azaborindenyl **27** (*14*). Formally **29** is constructed by the fusion of **8** and the aromatic 1,2-dihydro-1,2-azaborine (**26**) at the BN positions. (See Scheme 6). Although **25** is isoelectronic with indenyl, it is likely to be more electron rich (*15*).

The synthesis illustrated in Scheme 7 shows that boramine **28** can be constructed in a stepwise, straightforward fashion. The treatment of **28** with 5 mol % Grubbs catalyst in refluxing CH$_2$Cl$_2$ gives a 50% yield of **29** which is formed by a double RCM sequence. Dehydrogenation of **29** with DDQ in pentane affords a 30% yield of azaborindene **30** as a pale yellow liquid which has a strong indene-like odor. The sequential reaction of **30** with KN(SiMe$_3$)$_2$ followed by Cp* ZrCl$_3$ in toluene affords 60% of crystalline **31**. The molecular structure of **31**, illustrated in Figure 2, closely resembles that of the

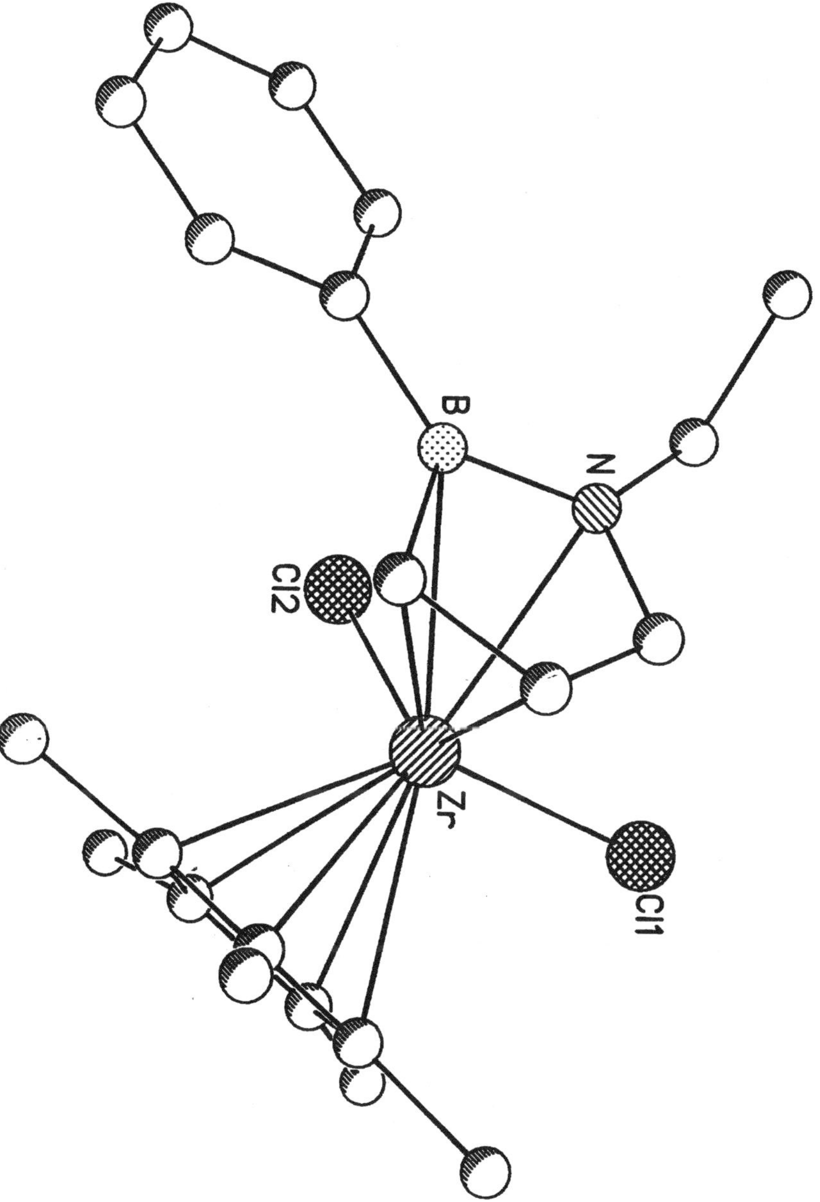

Figure 1. The molecular structure of (η^5-1-ethyl-2-phenyl-1,2-azaborolyl) (η^5-penta- methylcyclopentadienyl) zirconium dichloride (24).

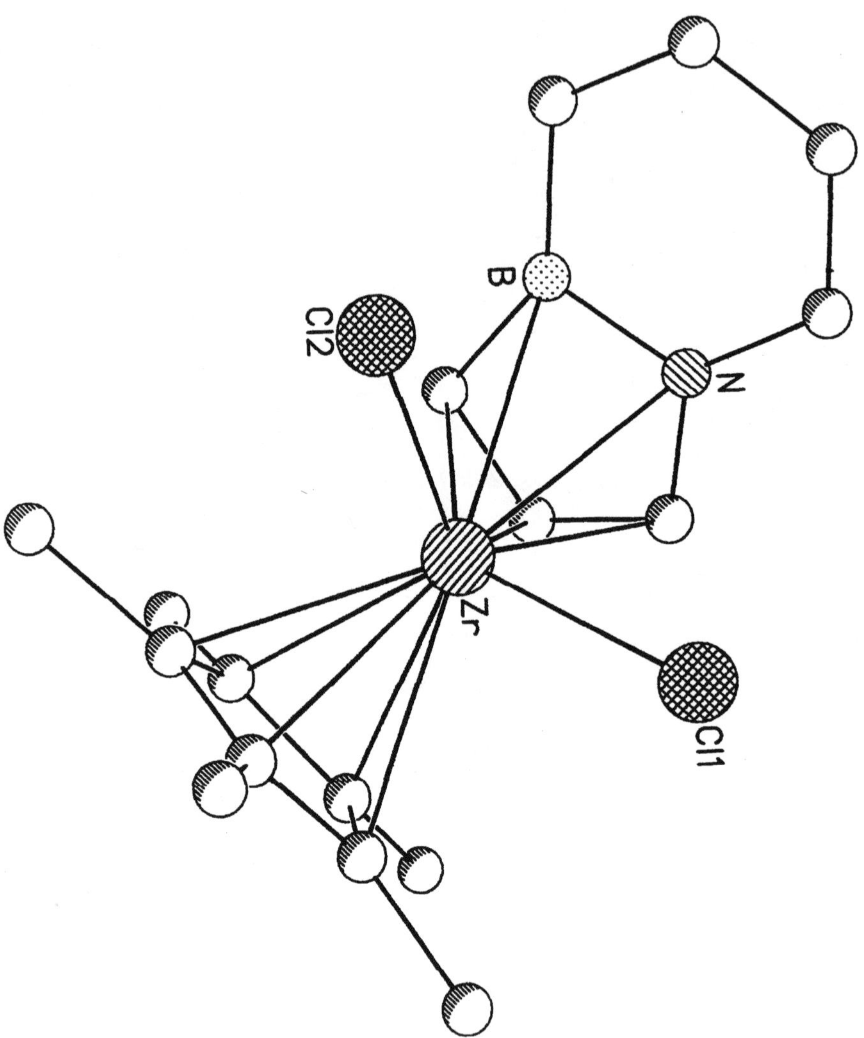

Figure 2. The molecular structure of (η^5-3a,7a-azaborindenyl) (η^5-pentamethyl-cyclopentadienyl) zirconium dichloride (29).

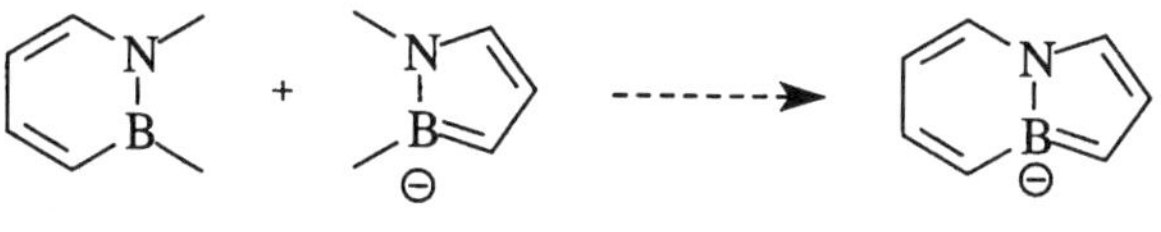

Scheme 6: The formal fusion of 26 and 8 at the BN positions gives 3a,7a-azaborindenyl 27

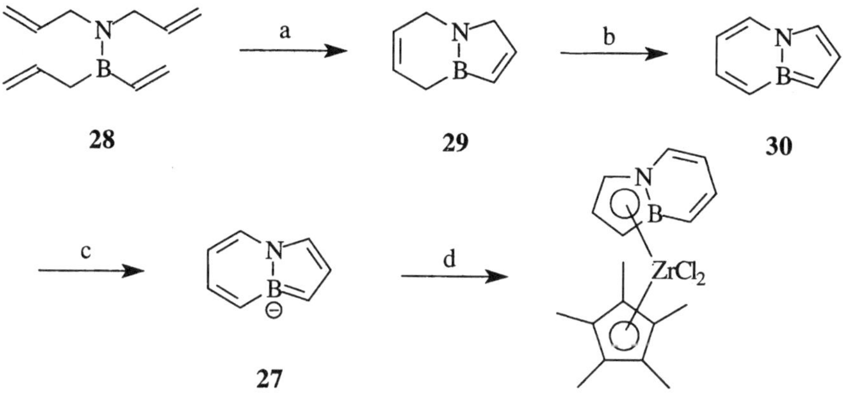

Scheme 7: Synthesis of 3a,7a-azaborindenyl

*Key: a) (PCy₃)₂(PhCH)RuCl₂, b) DDQ, c) (SiMe₃)₂NK, d) Cp*ZrCl₃*

corresponding indenyl complex Cp*(Ind)ZrCl₂ (*16*). Preliminary experiments indicate that MAO activated **25** and **31** are good ethylene polymerization catalysts.

Future Outlook

We have developed new anionic heterocyclic boron containing ligands which can be used as surrogates for Cp in Cp₂ZrCl₂. On activation by MAO in the usual manner these zirconium complexes form good catalysts for ethylene polymerization. An attractive future goal is to incorporate these Cp surrogates into complexes with geometries which have shown to form both highly active and stereoselective catalysts. Thus we are now attempting to prepare azaborindenyl analogs of the Brintzinger C_2-symmetry complexes **2**.

Acknowledgements

This work was supported by the National Science Foundation and the Dow Chemical Company.

References

1. Togni, A.; Halterman, R.L.; Eds *Metallocenes,* Vol 1,2. Wiley, New York, NY, 1998.

2. For recent reviews see:
   (a) Brintzinger, H.H.; Fischer, D.; Mülhaupt, R,; Rieger, B.; Waymouth, R.M. Angew. Chem. Int. Ed,; Engl. **1995**, *34*, 1143.
   (b) Fink, G.; Mülhaupt, R.; Brintzinger, H.H.; Eds. *Ziegler Catalysts*, Springer-Verlag, Berlin, Germany, 1995.
   (c) Janiak, C. "Metallocene Catalysts for Olefin Polymerization" in ref 1 vol 2, pp 547-623.
   (d) Gladysz, J.A. *Ed. Chem Rev.* **2000**, *100*, 1167-1682.

3. (a) Wild, R.R.W.P.; Zsolnai, L.; Huttner, G.; Brintzinger, H.H. *J. Orgnaomet. Chem.* **1982**, *232*, 233.
   (b) Wild, F.R.W.P.; Wasiucionek, M.; Huttner, G.; Brintzinger, H.H.; *J. Organomet. Chem.* **1985**, *288*, 63.
   (c) Kaminsky, W.; Külper, K.; Brintzinger, H.H.; Wild, F.W.R.P. *Angew. Chem. Int. Ed. Engl.* **1985**, *24*, 507.

4. Bazan, G.C.; Rodriguez, G.; Ashe, A.J., III; Al-Ahmad, S.; Müller, C. *J. Am. Chem. Soc,* **1996**, *118*, 2291.

5. Bazan, G.C.; Rodriguez, G.; Ashe, A.J., III; Al-Ahmad, S.; Kampf, J.W. *Organometallics* **1997**, *16*, 2492.

6. Ashe, A.J., III; Kampf, J.W.; Müller, C.; Schneider, M. *Organometallics* **1996**, *15*, 387.

7. (a) Schulze, J.; Schmid, G *Angew. Chem. Int. Ed. Engl.*, **1980**, *19*, 54.
   (b) Schmid, G. In *Comprehensive Heterocyclic Chemistry II*, Shinkai, I. Ed. Pergamon, Oxford, UK, **1996**, vol 3, p737.

8. Nagy, S.; Krishnamurti, R.; Etherton, B.P. PCI Int Appl. WO96 -34, 021; *Chem. Abstr.*, **1997**, *126*, 19432j.

9.   Ashe, A.J., III; Fang, X.; Kampf, J.W., *Organometallics,* **1999**, *18,* 1363.

10. Schiesher, M.W. *Ph.D. Dissertation* University of Michigan, 2002.

11. Ashe, A.J., III; Fang, X.; Kampf, J.W. *Organometallics,* **2000**, *19,* 4935.

12. Ashe, A.J., III; Fang, X. *Org. Lett.,* **2000**, *2,* 2089.

13. Ashe, A.J., III; Yang, H. Unpublished data

14. Ashe, A.J., III; Yang, H.; Fang, X.; Kampf, J.W. *Organometallics,* **2002**, *21,* 4578.

15. Liu, S.-Y.; Lo, M.M.-C.; Fu, G.C. *Angew. Chem. Int. Ed. Engl.,* **2002**, *41,* 174.

16. Schmid, M.A.; Alt, H.G.; Milius, W. *J. Organomet. Chem.,* **1996**, *514,* 45.

Chapter 3

FI Catalysts for Highly Controlled Living Polymerization of Ethylene and Propylene: Creation of Precisely Controlled Polymers

Makoto Mitani and Terunori Fujita

Research and Development Center, Mitsui Chemicals, Inc., 580–32 Nagaura, Sodegaura, Chiba 299–0265, Japan

The unique catalytic behavior of Ti complexes with fluorine-containing phenoxy–imine chelate ligands, fluorinated Ti-FI Catalysts, has been discussed. Fluorinated Ti-FI Catalysts with MAO activation initiate highly-controlled living polymerization of propylene as well as ethylene above room temperature and create a variety of precisely-controlled polymers previously unaccessible using Ziegler–Natta catalysts, such as high molecular weight monodisperse polyethylenes (M_n > 400,000), highly syndiotactic monodisperse polypropylenes with remarkably high T_ms (T_m 156 °C), and block copolymers comprised of crystalline and amorphous segments, with high efficiency.

Making polymers with perfect control over molecular weight, composition and architecture of the polymer has been a long-standing challenge for synthetic polymer chemists. Living olefin polymerization enables consecutive enchainment of monomer units without termination, and thus is a powerful tool for precisely controlling the molecular weight as well as the composition and architecture of the polymer. With living olefin polymerization, it is possible to prepare precisely-controlled polymers such as monodisperse polymers and block copolymers, which are difficult to produce from conventional Ziegler–Natta catalysis. In addition, living olefin polymerization permits the synthesis of end-functionalized polymers if special initiation and/or quenching methods are applied. Consequently, there has been great interest in the development of high performance catalysts for living olefin polymerization.

In general, living olefin polymerization requires low temperatures, normally below room temperature, to reduce the rates of chain termination and transfer reactions. Therefore, low activities and insufficient molecular weights of resultant polymers are typically observed. In addition, known catalysts promote the living polymerization of either ethylene or α-olefins such as propylene and 1-hexene but not both, resulting in the limited success in the synthesis of polyolefinic block copolymers. Moreover, only a few catalysts have been developed that initiate living polymerization with control of polymer stereochemistry. These drawbacks highly restrict the utility of the living olefin polymerization techniques for the preparation of the above mentioned precisely-controlled polymers especially that of block copolymers consisting of crystalline and amorphous segments.

Recent advances in the rational design of transition metal complexes for olefin polymerization have spurred the development of catalysts that promote living olefin polymerization at relatively high temperatures such as room temperature (*1*). Additionally, some catalysts have developed, which induce living polymerization of 1-hexene or other olefins (*2–8*) and stereospecific living polymerization of 1-hexene and vinylcyclohexane (*9, 10*). There are, however, a limited number of examples of catalysts that promote the room-temperature living polymerization of ethylene (*11–15*) and/or propylene (*16–20*). Moreover, highly stereospecific living polymerization of propylene has not yet been achieved despite the fact that nearly perfect stereospecific propylene polymerization can be performed with appropriately designed group 4 metallocene catalysts (*21*). Accordingly, there are many important goals still remaining in the field of living olefin polymerization.

In our own efforts to develop high performance catalysts for olefin polymerization, we have investigated well-defined transition metal complexes with non-symmetric [O, N], [N, N] or [O, O] chelate ligands having moderate electron donating properties. We found a new family of group 4 transition metal complexes featuring phenoxy-imine chelate ligands, named FI Catalysts, which exhibit high catalytic performance for the polymerization of ethylene and/or α-olefins including higher α-olefins and dienes (*22–26*).

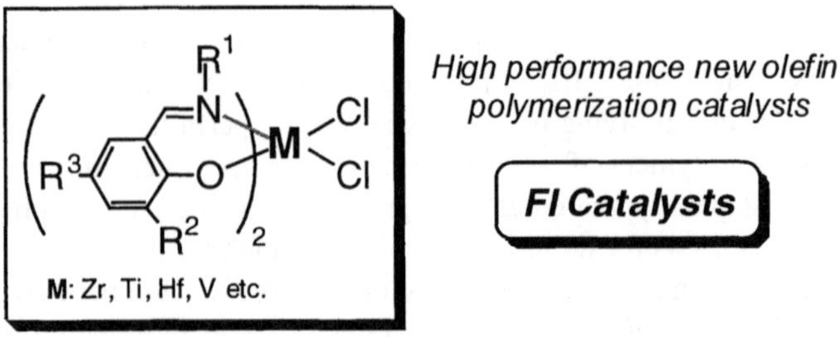

Figure 1. *General structure of FI Catalysts.*

For example, FI Catalysts are capable of producing vinyl-terminated low molecular weight polyethylenes (M_w 2,100), ultra-high molecular weight ethylene/propylene copolymers (M_w 10,200,000 vs PS standards), isotactic or syndiotactic polypropylenes, and high molecular weight atactic poly(1-hexene) including frequent regioerrors, with high efficiency. Further studies aimed at developing high performance FI Catalysts resulted in the discovery of a series of fluorinated Ti-FI Catalysts which exhibit unprecedented catalytic performance for the living polymerization of ethylene and/or propylene (*27–33*). In this article, we describe the catalytic behavior of fluorinated Ti-FI Catalysts in order to show the usefulness of this important class of catalysts to create precisely-controlled polymers which are unobtainable with conventional Ziegler–Natta catalysts.

Living Ethylene Polymerization

FI Catalysts possess ligands that are readily tailored synthetically from both an electronic and steric point of view, and thus have a wide range of catalyst design possibilities. In exploring new derivatives of FI Catalysts, with this advantage in mind, we have investigated FI Catalysts possessing heteroatom(s) and/or heteroatom-containing substituent(s) in the ligand. As a result, we discovered a Ti-FI Catalyst **1** having a perfluorophenyl at R^1 position, which exhibited unique catalytic performance for the living polymerization of ethylene (Table 1). FI Catalyst **1** with MAO activation at 50 °C under atmospheric pressure afforded high molecular weight polyethylene (M_n 412,000) having an extremely narrow M_w/M_n of 1.13, with exceptionally high activity (Turnover Frequency: TOF; 21,500 min^{-1}). The M_w/M_n value suggested that FI Catalyst **1**/MAO system may have the characteristics of a living ethylene polymerization under the given conditions.

Table 1. Results of Ethylene Polymerization with FI Catalyst 1/MAO

entry	catalyst	temp. (°C)	time (min)	polymer yield (g)	TOF (min⁻¹)	M_n[a] (×10³)	M_w/M_n[a]
1	1	25	0.5	0.149	21,200	191	1.15
2	1	25	1	0.283	20,200	412	1.13
3	1	50	0.5	0.172	24,500	257	1.08
4	1	50	1	0.302	21,500	424	1.13
5	Cp₂ZrCl₂	25	1	0.258	18,400	157	1.73
6	Cp₂ZrCl₂	50	1	0.433	30,900	136	2.26

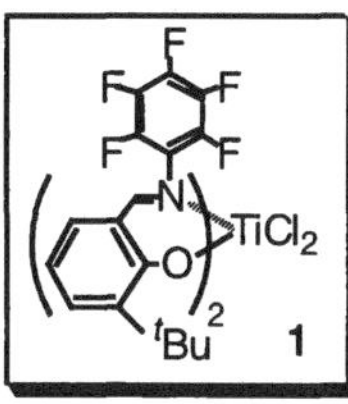

Conditions: cat. (0.5 μmol), cocat. MAO (1.25 mmol), 1 atm.
[a] Determined by GPC using polyethylene calibration.

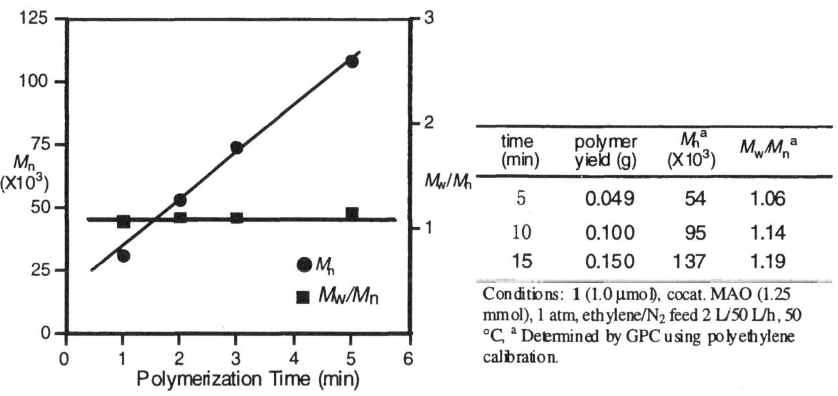

time (min)	polymer yield (g)	M_n[a] (×10³)	M_w/M_n[a]
5	0.049	54	1.06
10	0.100	95	1.14
15	0.150	137	1.19

Conditions: 1 (1.0 μmol), cocat. MAO (1.25 mmol), 1 atm, ethylene/N₂ feed 2 L/50 L/h, 50 °C, [a] Determined by GPC using polyethylene calibration.

Figure 2. *Plots of M_n and M_w/M_n as a function of polymerization time for ethylene polymerization with FI Catalyst 1/MAO at 50 °C.*

(Reproduced from reference 29. Copyright 2002 American Chemical Society.)

A plot of M_n and M_w/M_n values at 50 °C vs polymerization time is shown in Figure 2. The M_n value increased linearly with polymerization time, and the narrow M_w/M_n value (1.06–1.19) was retained for each run, indicating that the system is living. The molecular weight of the polyethylene is close to that calculated from the monomer/catalyst ratio obtained from the mass of the polyethylene produced, indicating that catalyst efficiency is substantially quantitative.

The M_n value, 412,000, represents one of the highest molecular weights among polyethylenes produced in a living manner. In addition, as far as we are aware, the TOF of 21,500 min⁻¹ is the highest activity reported to date with respect to living ethylene polymerization, and is comparable to those seen with early metallocenes such as Cp_2ZrCl_2 under the same conditions. Moreover, this is the first example of living ethylene polymerization at a high temperature of 50 °C. Considering that the MAO used as the cocatalyst is a potential chain transfer agent and hence usually living olefin polymerization can only be achieved using a borate cocatalyst instead of MAO, living ethylene polymerization with MAO exhibited by the system at 50 °C is highly significant. Notably, at 90 °C, 1/MAO produced polyethylene with a fairly narrow molecular weight distribution (M_w/M_n 1.30, M_n 167,000).

Since the corresponding non-fluorinated Ti-FI Catalyst 2/MAO system does not exhibit living ethylene polymerization and produces classical unimodal polyethylene having an M_w/M_n value of ca. 2 under identical conditions, we postulated that the fluorine atom(s) in the ligand is responsible for this highly-controlled living polymerization. The molecular structures of FI Catalysts 1 and 2 are presented in Figure 3. A comparison between the two structures revealed that 1 possesses a considerably wider Ti–N–C–C torsion angle involving the phenyl on the imine-nitrogen than 2, and thus ortho-fluorines of 1 are situated near Cl atoms. Since Cl-bound sites are potential olefin polymerization sites, ortho-fluorines might participate in the living polymerization. We investigated ethylene polymerization behavior of various fluorinated Ti-FI Catalysts 3–8. The results are collected in Table 2. There are clear differences in polymerization results depending on the fluorine atom substitution patterns.

1. Ti-FI Catalysts with ortho-fluorine(s) in R^1-phenyl group promote living ethylene polymerization. (The living nature of the polymerization for 3–5 was demonstrated by the linear relationship between M_n and polymerization time as well as narrow $M_w/M_n < 1.25$)
2. Activity displayed by an FI Catalyst increases with the increase in the number of fluorine atoms in both living- and non-living systems.
3. Activity displayed by a living-type FI Catalyst is substantially lower compared with that for a non-living type FI Catalyst having the same number of fluorine atoms.

It is reasonable that a more electron-withdrawing ligand would generate a more electrophilic Ti center. Therefore, result 2 provides a clear demonstration that for Ti-FI Catalysts the electrophilicity of the Ti center in active species plays a dominant role in determining the catalytic activity, unlike the group 4 metallocene catalysts whose activity is decreased by introducing electron-withdrawing substituent(s) in a Cp ligand. Result 1, namely the "livingness" observed for FI Catalysts having ortho-fluorine(s), shows that the presence of the ortho-fluorine(s) is a requirement for living polymerization. ^{13}C NMR as well as IR analysis revealed that the polyethylenes arising from the non-living type FI Catalysts possess vinyl end-groups (ca. 0.1 per 1000 carbon atoms),

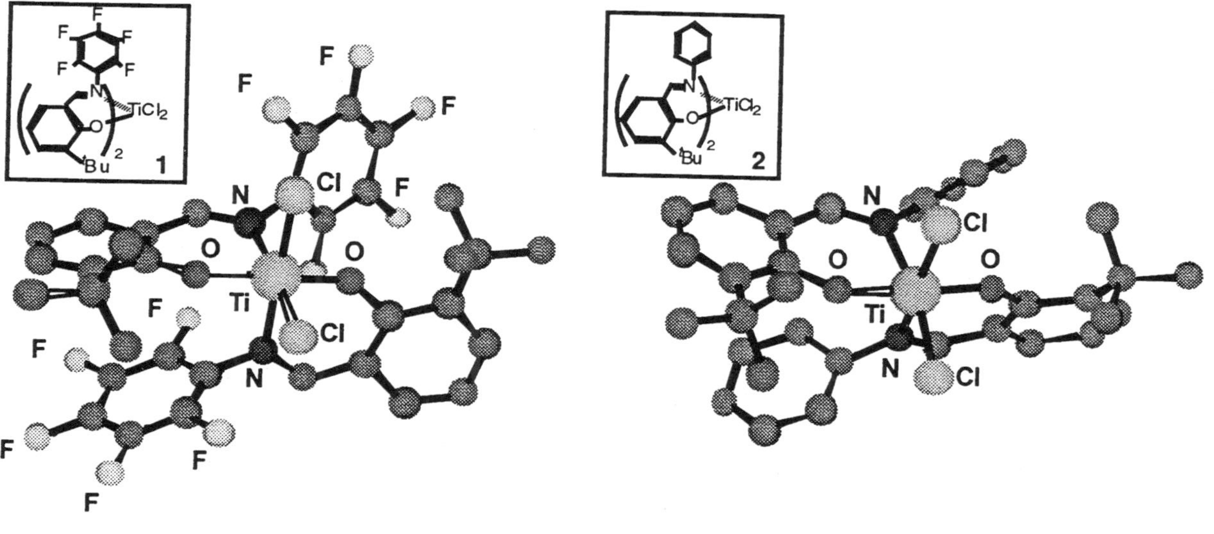

Figure 3. X-ray structures of FI Catalyst 1 and 2.

Table 2. Ethylene Polymerization Results with FI Catalysts 3–8/MAO

entry	catalyst	polymer yield (g)	TOF (min^{-1})	$M_n{}^a$ ($\times 10^3$)	$M_w/M_n{}^a$
1	3	0.081	1,440	145	1.25
2	4	0.069	492	64	1.05
3	5	0.053	76	13	1.06
4	6	0.372	26,500	98	1.99
5	7	0.267	19,000	129	1.78
6	8	0.177	3,160	128	2.18

Conditions: cat. (0.4 µmol entries 1, 6; 1.0 µmol entry 2; 5.0 µmol entry 3; 0.5 µmol entries 4, 5), time (5 min entries 1, 2, 3, 6; 1 min entries 4, 5), cocat. MAO (1.25 mmol), 50 °C, 1 atm, ethylene feed 100 L/h, toluene 250 mL. [a] Determined by GPC using polyethylene calibration.

complex	1	2	3	4

complex	5	6	7	8

Source: Reproduced from reference 29. Copyright 2002 American Chemical Society.

indicating that β-hydrogen transfer is the main termination pathway for these systems. These results suggest that the ortho-fluorine suppresses the β-hydrogen transfer. As reported, DFT calculations are an effective tool for analyzing the structure of bis(phenoxy–imine)group 4 metal complexes. Therefore, the calculations were performed on active species generated from FI Catalysts **1**, and **3–5** (n-propyl; model for polymer chain) to elucidate the role of the ortho-fluorine for the living polymerization (Table 3).

It was demonstrated that the ortho-fluorine and a β-hydrogen of a polymer chain are located at a distance of 2.276 to 2.362 Å, a distance which is well within the range of non-bonding interactions. Additionally, the C-H$_\beta$ bond elongation (1.113 Å) was suggested by the calculations, which is probably the result of an interaction between the ortho-fluorine with the β-hydrogen. On the basis of the calculation results, the interaction is mainly governed by electrostatic energy, estimated approximately to be -30 kJ/mol, between the negatively charged fluorine and the positively charged β-hydrogen. The electrostatic energy is large enough to mitigate the β-hydrogen transfer to the Ti metal and/or a reacting monomer. Besides, the transition state leading to the β-hydrogen transfer is probably disfavored by the fact that the β-hydrogen is positively charged and stabilized by the ortho-fluorine though the β-hydrogen should behave as a hydride in the β-hydrogen transfer process. These facts result in the prevention of the β-hydrogen transfer. Interestingly, FI Catalyst **9**, the chlorinated counterpart of fluorinated FI Catalyst **5**, with MAO activation at 25 °C provided polyethylene having a narrow molecular weight distribution (M_w/M_n 1.23), implying that the interaction with a β-hydrogen is potentially achieved by any substituent having lone-pair electrons. The interaction provides a new breakthrough for achieving highly-controlled living ethylene polymerization, and may be extended to the living polymerization of propylene since a growing polypropylene chain has a β-hydrogen.

Living Propylene Polymerization

X-ray crystallographic analysis established that Ti-FI Catalyst **1** possesses C_2 symmetry, suggesting that **1** may combine living enchainment with control of polymer stereochemistry. Propylene polymerization using **1**/MAO under atmospheric pressure at 25 °C yielded crystalline polypropylene, M_n 28,500, having an extremely narrow M_w/M_n of 1.11, with a TOF of 87 h^{-1} (Table 4).

The living nature of the polymerization was confirmed by the linear relationship between M_n and polymerization time as well as narrow M_w/M_n values for all runs (Table 4, entries 3–6). The polypropylene arising from **1**/MAO exhibits a peak melting temperature (T_m) of 137 °C, indicative of a high degree of stereocontrol. To our surprise, microstructural analysis by ^{13}C NMR

Table 3. F–H_β Interactions Calculated by DFT

R:	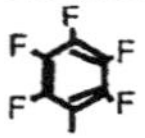 $F{\cdot}F{\cdot}F{\cdot}F$	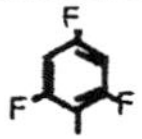 $F{\cdot}F$	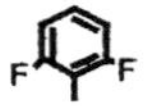 $F{\cdot}F$	 F
catalyst	1	3	4	5
$r(F–H_\beta)^a$	2.276	2.362	2.346	2.324
$q(F)^b$	-0.466	-0.470	-0.476	-0.482
$q(H_\beta)^c$	0.095	0.108	0.111	0.117
$ES(F–H_\beta)^d$	-27.1	-29.9	-31.2	-33.6

[a] F–H_β distance (Å). [b] Mulliken charge of the nearest o-F to H_β. [c] Mulliken charge of H_β. [d] Electrostatic energy for F–H_β interaction (kJ/mol).

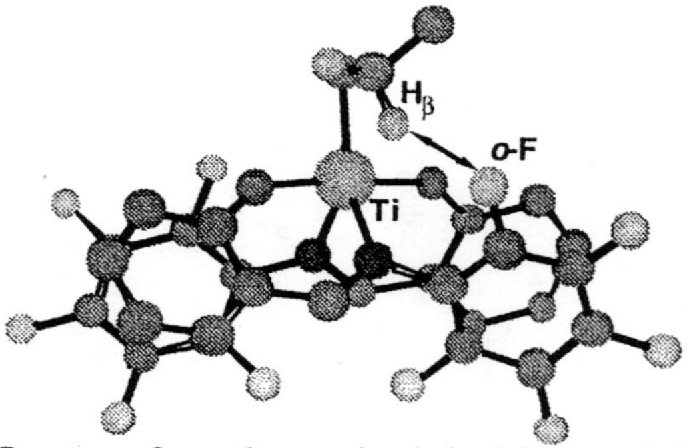

Structure of an active species derived from catalyst **1** calculated by DFT. ^{t}Bu groups are omitted for clarity.

Source: Reproduced from reference 29. Copyright 2002 American Chemical Society.

revealed that the monodisperse crystalline polymer formed with 1/MAO is highly syndiotactic polypropylene containing 87 % rr triads (Figure 4). The syndiotactic polypropylene involves isolated m-dyad errors (rrrm and rmrr) in the methyl pentad region, suggesting that the syndiospecific polymerization proceeds via a chain-end control mechanism.

Table 4. Results of Propylene Polymerization Using Catalyst 1/MAO

entry	Press. (MPa)	time (h)	polymer yield (g)	TOF (h^{-1})	M_n^a $(\times 10^3)$	M_w/M_n^a	T_m^b (°C)
1	0.1	3	0.096	76	20.5	1.09	137
2	0.1	5	0.183	87	28.5	1.11	137
3	0.6	1	0.158	375	30.9	1.07	135
4	0.6	2	0.312	371	52.8	1.10	–
5	0.6	3	0.460	364	73.8	1.10	135
6	0.6	5	0.713	339	108.0	1.14	135

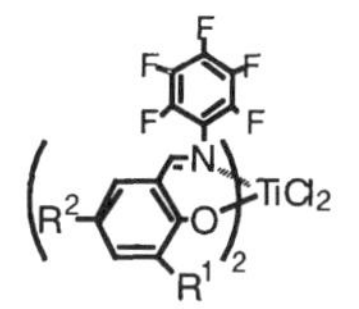

1 ; $R^1 = {}^tBu$, $R^2 = H$
10; $R^1 = {}^tBu$, $R^2 = {}^tBu$
11; $R^1 = SiMe_3$, $R^2 = H$
12; $R^1 = H$, $R^2 = H$

Conditions: cat. (10 μmol), cocat. MAO (2.5 mmol), 25 °C.
[a] Determined by GPC using polypropylene calibration.
[b] Melting temperature measured by DSC.

FI Catalyst **1** is unique in its ability to generate a highly stereoregular polymer through a chain-end control mechanism at room temperature. In fact, the rr triad, 87%, represents one of the highest values reported to date for syndiotactic polypropylenes produced via a chain-end control mechanism. Also, the rr triad is among the highest reported for monodisperse syndiotactic polypropylenes. To the best of our knowledge, FI Catalyst **1** is the first example of a living and, at the same time, highly stereoselective catalyst for propylene polymerization. In addition, FI Catalyst **1** is the first example of living polymerization catalyst capable of initiating the living polymerization of propylene as well as ethylene, suggesting that **1** is well-suited for the synthesis of block copolymers from ethylene and propylene.

Based on combinatorial methods Coates et al. concurrently obtained Ti-FI Catalyst **10** (*34*), having an additional *tert*-butyl group in the phenoxy-benzene ring, which shows similar catalytic performance to FI Catalyst **1** for living propylene polymerization (**1**; TOF 87 h^{-1}, M_n 28,500, M_w/M_n 1.11, T_m 137 °C: **10**; TOF 104 h^{-1}, M_n 29,800, M_w/M_n 1.15, T_m 135 °C: 25 °C, atmospheric pressure, 5h). These results indicate that *tert*-butyl group at the R^2 position has no significant influence on the catalytic performance of fluorinated Ti-FI Catalysts for propylene polymerization.

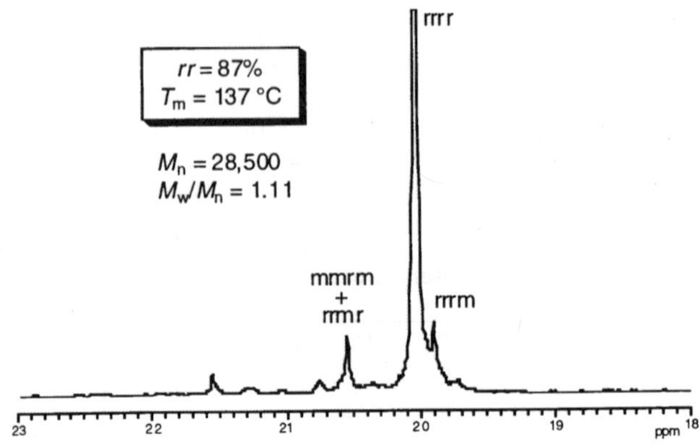

Figure 4. *^{13}C NMR spectra of monodisperse syn-PP produced by 1.*

Chain-end analysis using ^{13}C NMR spectroscopy revealed that the living syndiotactic polypropylene having an M_n of 2,000 contains three kinds of chain-end groups out of seven possible chain-end groups (Scheme 1); isopentyl (32.6%), an initiation chain-end, *n*-propyl (34.8%), a termination chain-end, and isobutyl (32.6%), an initiation or termination chain-end, (Figure 5). If the first propylene is enchained by a 2,1-insertion, the resulting polymer possesses an ethyl chain-end group. However, no ethyl chain-end group is found in the syndiotactic polypropylene. Thus, the first propylene is inserted into the Ti-Me bond of an initial active species originating from **1** and MAO in a 1,2-fashion. The *n*-propyl chain-end group, 34.8%, is the termination chain-end derived from the termination after consecutive 2,1-, 2,1- insertion, indicating that 2,1-insertion is predominant for chain propagation (ca. 70%). Moreover, the living syndiotactic polypropylenes having M_n of 3,300 (M_w/M_n 1.08) and 9,700 (M_w/M_n 1.05) possess isopentyl, *n*-propyl, and isobutyl chain-end groups with practically the same content as the living polypropylene discussed above (M_n 2,000). Therefore, the syndiospecific propylene polymerization exhibited by Ti-FI Catalyst **1** is initiated exclusively via 1,2-insertion followed by 2,1-insertion as the principal mode of polymerization.

As far as we are aware, this is the first example of a predominant 2,1-insertion mechanism for chain propagation displayed by a group 4 metal based catalyst. Predominant 2,1-insertion may be derived from the steric congestion near the polymerization reaction center due to a polymer chain and the ligand, which places the methyl of the reacting propylene opposite direction from a polymer chain. Recently, the generality of this highly unusual mechanism of insertion in propylene polymerization with Ti-FI Catalysts has been confirmed by other research groups (*35–37*).

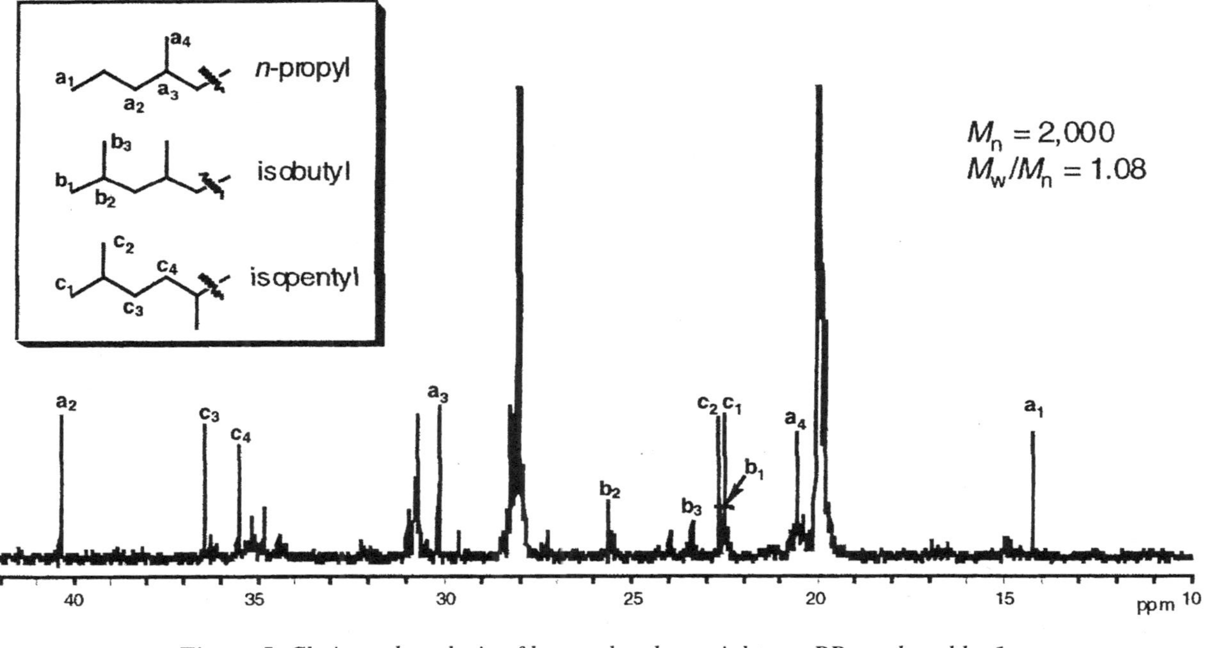

Figure 5. Chain-end analysis of low molecular weight syn-PP produced by 1.

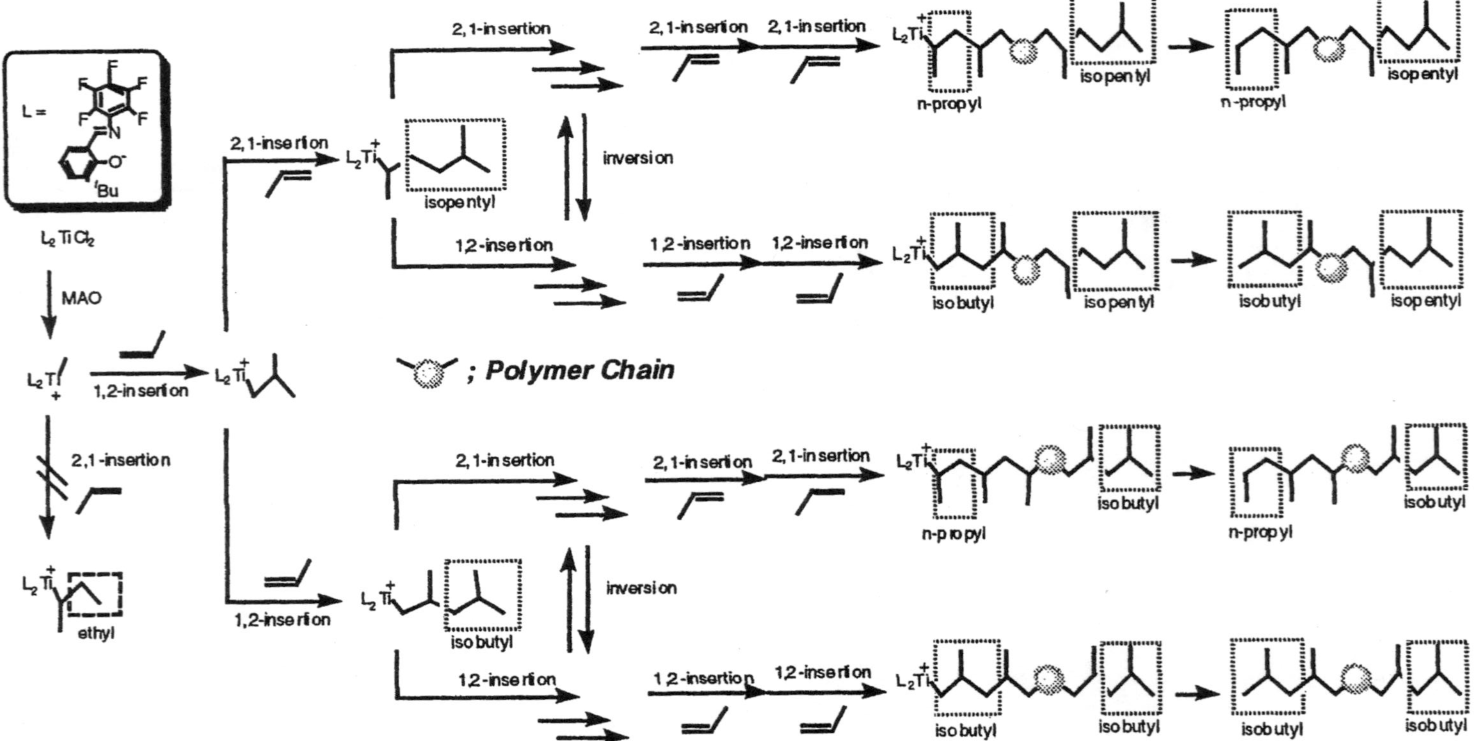

Scheme 1. Mechanism of initiation and chain propagation for propylene polymerization using catalyst 1/MAO.

Further efforts aimed at developing higher performance FI Catalysts for living propylene polymerization led to the discovery of FI Catalyst **11** which provides much higher tacticity polymers with extremely high T_ms via a chain-end control mechanism. Thus, FI Catalyst **11** having a trimethylsilyl at the R^1 position yielded (at 25 °C) highly syndiotactic monodisperse polypropylene (*rr* 93%, M_n 47,000, M_w/M_n 1.08) with a very high T_m (152 °C). Such an extremely high level of chain-end control is unprecedented in a propylene polymerization. The syndiotactic polypropylene produced by **11** at 0 °C displayed an exceptionally high T_m of 156 °C (*rr* 94 %) (*38*), which represents the highest T_m value among syndiotactic polypropylenes ever known (*39, 40*).

Surprisingly, at 50 °C, **11** furnished monodisperse syndiotactic polypropylene with a very high T_m of 150 °C. This is significant because generally chain-end control rapidly loses its stereoregulating ability at elevated temperatures. As a result of the interaction between the ortho-fluorine and a β-hydrogen, a polymer chain is possibly forced to be oriented in one direction and fixed firmly in one conformation. The rigid conformation of the polymer chain near the polymerization reaction center may directly and/or indirectly help the stereochemical information of the α-carbon transfer to the incoming propylene monomer, resulting in highly syndiospecific polymerization even at elevated temperatures.

To gain further information on the effect of R^1 substituent, we synthesized FI Catalyst **12** having a hydrogen atom at the R^1 position and investigated its potential as propylene polymerization catalyst. FI Catalyst **12** showed considerably higher activity than FI Catalysts **1**, **10** and **11**, probably due to its structurally open nature, and much lower syndiospecificity (*rr* 43%, no T_m). These results suggest that the steric bulk of the R^1 substituent plays a pivotal role in determining the syndiospecificity of the polymerization. It is clear that the ligand structure of the FI Catalyst has a dramatic effect on polymerization behavior, and that sterically bulky substituent results in highly stereospecific, thermally robust chain-end controlled propylene polymerization. Therefore, we have given the name *"ligand-directed chain-end control"* to this highly-controlled propylene polymerization.

Block Copolymer Formation

An attractive feature of living olefin polymerization catalysts is their ability to generate polyolefinic block copolymers, which may be applied to a broad spectrum of applications including compatibilizers, elastomers, and composite materials. As discussed, fluorinated Ti-FI Catalysts are capable of promoting the living polymerization of propylene as well as ethylene to give high molecular weight monodisperse polyethylenes (M_n > 400,000) and syndiotactic polypropylenes (M_n > 100,000). In addition, high molecular weight monodisperse ethylene/propylene copolymers (M_n > 80,000, M_w/M_n 1.07–1.13) having a wide range of propylene content (15–48 mol %) were synthesized using catalyst **1**/MAO system at 25 °C. These results indicated that with fluorinated Ti-FI Catalysts it is possible to prepare polyolefin block copolymers from ethylene, propylene, and/or ethylene/propylene. Thus, a number of unique block copolymers have been synthesized using catalyst **1** with MAO system (Table 5).

For instance, a new A-B diblock copolymer consisting of crystalline and amorphous segments, PE-*b*-poly(ethylene-*co*-propylene), was prepared from ethylene and ethylene/propylene. Addition of **1**/MAO to an ethylene saturated toluene at 25 °C resulted in the rapid formation of the first polyethylene block segment (M_n 115,000, M_w/M_n 1.10). The ethylene/propylene (1/3) feed formed a sequential poly(ethylene-*co*-propylene) segment. The resulting high molecular weight PE-b-poly(ethylene-co-propylene) block copolymer, (M_n 211,000, M_w/M_n 1.16) possessed a propylene content of 6.4 mol %. Similarly, PE-*b*-sPP, sPP-*b*-poly(ethylene-*co*-propylene), PE-*b*-poly(ethylene-*co*-propylene)-*b*-sPP, and PE-*b*-poly(ethylene-*co*-propylene)-b-PE block copolymers, previously unavailable with conventional Ziegler–Natta catalysts, were also prepared with **1**/MAO system by the sequential addition of the corresponding monomers. Thus, the usefulness of FI Catalysts/MAO systems for the synthesis of a wide array of block copolymers has been demonstrated. This powerful strategy for block cppolymer formation is further being applied to the preparation of block copolymers having novel architectures. Coates and co-workers have obtained a sPP-*b*-poly(ethylene-*co*-propylene) block copolymer having a narrow molecular weight distribution (M_n 145,100, M_w/M_n 1.12) with FI Catalyst **10** at 0 °C (*34*).

The block copolymers produced by FI Catalysts exhibit peak melting temperatures (T_m) of 123 °C for PE-*b*-poly(ethylene-*co*-propylene), 131 °C for PE-*b*-sPP, 127 °C for sPP-*b*-poly(ethylene-*co*-propylene), 123 °C for PE-*b*-poly(ethylene-*co*-propylene)-*b*-sPP, and 120 °C for PE-*b*-poly(ethylene-*co*-propylene)-*b*-PE, the T_m value being lower than the corresponding PE (133 °C) and sPP (137 °C). The decrease in T_m of the block copolymers is probably due to the fact that the crystalline state is disturbed by the segregation process of incompatible components comprised of crystalline sPP and/or PE and amorphous poly(ethylene-co-propylene) segments which are chemically linked, further confirming the formation of the block copolymers.

Transmission electron microscopy (TEM) micrograph of the press-sheet made of the sPP-*b*-poly(ethylene-*co*-propylene) formed with **1**/MAO system shows well-defined morphology of micro-phase separation (Figure 6). In Figure 6(A), white domains which correpond to the sPP segments form a very fine and uniform nanostructures, whereas those for the blend polymer (Figure 6(B); blend conditions; toluene, 100 °C, 1h) exhibit coarse and non-uniform structures. The block copolymer may possess a high potential as a compatibilizer.

Table 5. Synthesis of Various Block Copolymers

	prepolymer			diblock copolymer				triblock copolymer			
entry	1st block	$M_n{}^a$, $M_w/M_n{}^a$	$T_m{}^b$ (°C)	2nd block	$M_n{}^a$, $M_w/M_n{}^a$	propylene^c contents	$T_m{}^b$ (°C)	3rd block	$M_n{}^a$, $M_w/M_n{}^a$	propylene^c contents	$T_m{}^b$ (°C)
1	PE^d	115 000, 1.10	133	PP^e	136 000, 1.15	16.1	131	–	–	–	–
2	↑	↑	↑	E/P^f	211 000, 1.16	6.4	123	–	–	–	–
3	↑	↑	↑	↑	↑	↑	↑	PP	235 000, 1.15	14.1	123
4	↑	↑	↑	↑	↑	↑	↑	PE	272 000, 1.14	6.6	120
5	PP	27 000, 1.13	137	E/P	161 000, 1.51	40.3	127	–	–	–	–

[a] Determined by GPC using polypropylene calibration. [b] Melting temperature measured by DSC. [c] Total propylene contents (mol%). Determined by NMR. [d] Polyethylene. [e] Polypropylene. [f] Ethylene-propylene random copolymer.

Source: Reproduced from reference 29. Copyright 2002 American Chemical Society.

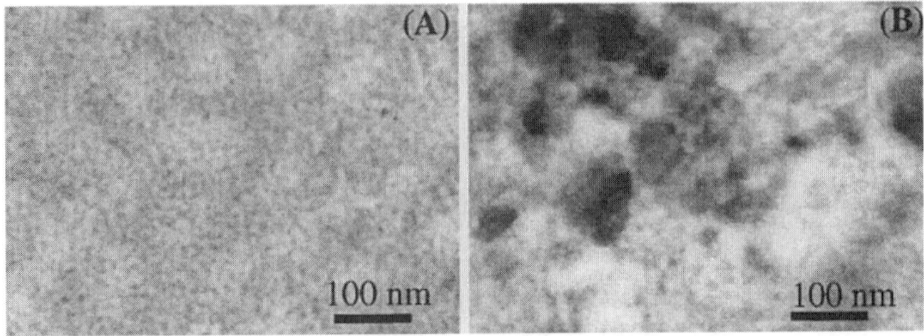

Figure 6. TEM micrographs of (A) sPP-b-poly(ethylene-co-propylene) and (B) sPP and poly(ethylene-co-propylene) blend polymer.

The catalytic performance of fluorinated Ti-FI Catalysts for the polymerization of ethylene and/or propylene has been summarized. These catalysts with MAO are capable of promoting highly-controlled living ethylene polymerization by the unprecedented interaction of the ortho-fluorine and a β-hydrogen of a growing polymer chain. This remarkable ortho-fluorine effect probably represents a novel strategy for the design of a new transition metal complex for living olefin polymerization. Fluorinated Ti-FI Catalysts initiate living, stereospecific propylene polymerization to produce highly syndiotactic monodisperse polypropylenes with very high T_ms via *ligand-directed chain-end control mechanism.* With chain-end group analysis, we have discovered a highly unusual and unprecedented propylene insertion mechanism in which the polymerization is initiated exclusively via 1,2-insertion followed by 2,1-insertion as the principal mode of polymerization. These results demonstrate that Ti-FI Catalysts have provided a new breakthrough for accomplishing highly-controlled living and/or highly-stereospecific polymerization of olefins. With fluorinated Ti-FI Catalysts, it is possible to prepare a number of unique polymers previously unavailable with conventional Ziegler–Natta catalysts, such as high molecular weight monodisperse polyethylenes and ethylene/propylene copolymers, highly syndiotactic monodisperse polypropylenes with extremely high T_ms, and a variety of polyolefinic block copolymers from ethylene and propylene. The results introduced herein along with our previous reports suggest that FI Catalysts have been widening the scope of polyolefinic materials accessible by transition metal-based catalytic technology.

Acknowledgment

We would like to thank Dr. M. Mullins for his fruitful discussions and suggestions. We are also grateful to J. Mohri, R. Furuyama, J. Saito, Y.

Yoshida, S. Ishii, H. Terao, K. Tsuru, Y. Suzuki, N. Matsukawa, S. Matsuura, Y. Inoue, S. Matsui, Y. Takagi, Y. Nakayama, H. Bando, Y. Tohi, H. Makio, T. Nakano, H. Tanaka, T. Matsugi, S. Kojoh for their research and technical assistance. We would also like to thank M. Onda for NMR analysis and T. Abiru for GPC analysis.

References

1. For a recent review see; Coates, G. W.; Hustad, P. D.; Reinarrtz, S. *Angew. Chem., Int. Ed.* **2002**, *41*, 2236–2257.
2. Scollard, J. D.; McConville, D. H. *J. Am. Chem. Soc.* **1996**, *118*, 10008–10009.
3. Baumann, R.; Stumpf, R.; Davis, W. M.; Liang, L. C.; Schrock, R. R. *J. Am. Chem. Soc.* **1999**, *121*, 7822–7836.
4. Schrodi, Y.; Schrock, R. R.; Bonitatebus, P. J. *Organometallics* **2001**, *20*, 3560–3573.
5. Mehrkhodavandi, P.; Schrock, R. R. *J. Am. Chem. Soc.* **2001**, *123*, 10746–10747.
6. Tshuva, E. Y.; Goldberg, I.; Kol, M.; Goldschmidt, Z. *Chem. Commun,* **2001**, 2120–2121.
7. Jeon, Y. M.; Park, S. J.; Heo, J.; Kim, K. *Organometallics* **1998**, *17*, 3161–3163.
8. Yoshida, Y.; Saito, J.; Mitani, M.; Takagi, Y.; Matsui, Ishii, S.; S.; Nakano, T.; Kashiwa, N.; Fujita, T. *Chem. Commun,* **2002**, 1298–1299.
9. Jayaratne, K. C.; Keaton, R. J.; Henningsen, D. A.; Sita, L. R. *J. Am. Chem. Soc.* **2000**, *122*, 10490–10491.
10. Jayaratne, K. C.; Sita, L. R. *J. Am. Chem. Soc.* **2001**, *123*, 10754–10755.
11. Yasuda, H.; Furo, M.; Yamamoto, H.; Nakamura, A.; Miyake, S.; Kibino, N. Macromolecules **1992**, 25, 5115–5116.
12. Mashima, K.; Fujikawa, S.; Nakamura, A. *J. Am. Chem. Soc.* **1993**, *115*, 10990–10991.
13. Brookhart, M.; DeSimone, J. M.; Grant, B. E.; Tanner, M. J. *Macromolecules* **1995**, *28*, 5378–5380.
14. Gottfried, A. C.; Brookhart, M. *Macromolecules* **2001**, *34*, 1140–1142.
15. Matsugi, T.; Matsui, S.; Kojoh, S.; Takagi, Y.; Inoue, Y.; Fujita, T.; Kashiwa, N. *Macromolecules* **2002**, *35*, 4880–4887.
16. Doi, Y.; Tokuhiro, N.; Soga, K. *Makromol. Chem., Macromol. Chem. Phys.* **1989**, *190*, 643–651 and references therein.
17. Killian, C. M.; Tempel, D. J.; Johnson, L. K.; Brookhart, M. *J. Am. Chem. Soc.* **1996**, *118*, 11664–11665.
18. Hagimoto, H; Shiono, T.; Ikeda, T. *Macromolecules* **2002**, *35*, 5744–5745.
19. Fukui, Y.; Murata, M.; Soga, K. *Macromol. Rapid Commun.* **1999**, *20*, 637–640.

20. Turner, H. W.; Hlatky, G. G. Exxon, PCT International Application 9112285, 1991 *(Chem. Abstr.* **1992**, *116*, 61330t).

21. Coates, G. W. *J. Chem. Soc., Dalton Trans.*, **2002**, 467–475.

22. Fujita, T.; Tohi, Y.; Mitani, M.; Matsui, S.; Saito, J.; Nitabaru, M.; Sugi, K.; Makio, H.; Tsutsui, T. Europe Patent, EP-0874005, 1998 *(Chem. Abstr.* **1998**, *129*, 331166).

23. Matsui, S; Tohi, Y.; Mitani, M.; Saito, J.; Makio, H.; Tanaka, H.; Nitabaru, M.; Nakano, T.; Fujita, T. *Chem. Lett.* **1999**, 1065–1066.

24. Matsukawa, N.; Matsui, S.; Mitani, M.; Saito, J.; Tsuru, K.; Kashiwa, N.; Fujita, T. *J. Mol. Catal. A* **2001**, *169*, 99–104.

25. Matsui, S.; Mitani, M.; Saito, J.; Tohi, Y.; Makio, H.; Matsukawa, N.; Takagi, Y.; Tsuru, K.; Nitabaru, M.; Nakano, T.; Tanaka, H.; Kashiwa, N.; Fujita, T. *J. Am. Chem. Soc.* **2001**, *123*, 6847–6856.

26. Recently, FI Catalysts and related complexes have been actively studied because of their high potential for olefin polymerization. For a recent review see; Makio, M.; Kashiwa, N.; Fujita, T. *Adv. Synth Catal.* **2002**, *344*, 477–493.

27. Mitani, M.; Yoshida, Y.; Mohri, J.; Tsuru, K.; Ishii, S.; Kojoh, S.; Matsugi, T.; Saito, J.; Matsukawa, N.; Matsui, S.; Nakano, T.; Tanaka, H.; Kashiwa, N.; Fujita, T. WO Pat. 01/55231 A1, 2001.

28. Saito, J.; Mitani, M.; Mohri, J.; Yoshida, Y.; Matsui, S.; Ishii, S.; Kojoh, S.; Kashiwa, N.; Fujita, T. *Angew. Chem. Int. Ed.* **2001**, *40*, 2918–2920.

29. Mitani, M.; Mohri, J.; Yoshida, Y.; Saito, J.; Ishii, S.; Tsuru, K.; Matsui, S.; Furuyama, R.; Nakano, T.; Tanaka, H.; Kojoh, S.; Matsugi, T.; Kashiwa, N.; Fujita, T. *J. Am. Chem. Soc.* **2002**, *124*, 3327–3336.

30. Saito, J.; Mitani, M.; Mohri, J.; Ishii, S.; Yoshida, Y.; Matsugi, T.; Kojoh, S.; Kashiwa, N.; Fujita, T. *Chem. Lett.* **2001**, 576–577.

31. Kojoh, S.; Matsugi, T.; Saito, J.; Mitani, M.; Fujita, T.; Kashiwa, N. *Chem. Lett.* **2001**, 822–823.

32. Saito, J.; Mitani, M.; Onda, M.; Mohri, J.; Ishii, S.; Yoshida, Y.; Nakano, T.; Tanaka, H.; Matsugi, T.; Kojoh, S.; Kashiwa, N.; Fujita, T. *Macromol. Rapid Commun.* **2001**, *22*, 1072–1075.

33. Mitani, M.; Furuyama, R.; Mohri, J.; Saito, J.; Ishii, S.; Terao, H.; Kashiwa, N.; Fujita, T. *J. Am. Chem. Soc.* **2002**, *124*, 7888–7889.

34. Hustad, P. D.; Tian, J.; Coates, G. W. *J. Am. Chem. Soc.* **2001**, *123*, 5134–5135.

35. Tian, J.; Hustad, P. D.; Coates, G. W. *J. Am. Chem. Soc.* **2002**, *124*, 3614–3621.

36. Lamberti, M.; Pappalardo, D.; Zambelli, A.; Pellecchia, C. *Macromolecules* **2002**, *35*, 658–663.

37. Kakugo et al. discovered that a sulfur-bridged bisphenoxy Ti complex forms nearly completely regioirregular PPs. (ref. Miyatake, T.; Mizumuma, K.; Kakugo, M *Makromol. Chem. Macromol. Symp.* **1993**, *66*, 203–214) It seems that phenoxy-based Ti complexes have a propensity to produce regioireregular polymers.

38. Obviously, the T_ms of sPP arising from fluorinated Ti-FI Catalysts are higher than the values expected from syndiotacticity of the polymers. The difference probably originates from the microstructures and/or unimodality of the polymers. For example; sPP formed with Ph$_2$C(Cp, 2,7-di-tert-Bu-Flu)ZrCl$_2$, [rr] 95%, T_m 148 °C; under the same DSC measurement conditions. Shiomura, T.; Kohno, M.; Inoue, N.; Yokote, Y.; Akiyama, M.; Asanuma, T.; Sugimoto, R.; Kimura, S.; Abe, M. Kodansha and Elsevier *Catalyst Design for Tailor-Made Polyolefins*, Tokyo and Amsterdam, **1994**, 327–338.

39. Veghini, D.; Henling, L. M.; Burkhardt, T. J.; Bercaw, J. E. *J. Am. Chem. Soc.* **1999**, *121*, 564–573.

40. Grisi, F.; Longo, P.; Zambelli, A.; Ewen, J. A. *J. Mol. Catal. A* **1999**, *140*, 22

Group 4 Octahedral Complexes Catalyzed the Stereoregular and Elastomeric Polymerization of Propylene

Victoria Volkis[1], Michal Shmulinson[1], Ella Shaviv[1],
Anatoli Lisovskii[1], Dorit Plat[1], Olaf Kühl[2], Thomas Koch[2],
Evamarie Hey-Hawkins[2], and Moris S. Eisen[1,*]

[1]Department of Chemistry and Institute of Catalysis Science and
Technology, Technion-Israel Institute of Technology, Haifa 32000, Israel
[2]Institute für Anorganishe Chemie der Universität at Leipzig,
D–04103 Leipzig, Germany

Some group 4 octahedral complexes with different ligations (benzamidinate, acetylacetonate, β-diketoimidinate, phosphinoamide) have been synthesized. We demonstrate that early transition metals with a *cis*-octahedral geometry, when activated by methylalumoxane, catalyzed the stereoregular polymerization of propylene, which can be modulated by pressure. The mechanism of formation and structure of the new type of elastomeric polypropylene obtained is discussed.

Introduction

During the last two decades a great advancement have been made in the synthesis and application of well-defined "single site" metallocene catalysts for the polymerization of α-olefins (*1-5*). Metallocene catalysts generally belong to the group 3 and 4 derivatives with two cyclopentadiene (Cp) rings, although some of these system (constrain geometry) contain one Cp and one pendant ligand (*6,7*). The structure of the ligands affect both electronic and sterically the properties of the metallocenes and thus predictably determined the stereospecificity of these catalysts (*8,9*). Recently, a great interest have been directed towards the synthesis of non-Cp complexes as potential catalysts for the polymerization of olefins as alternatives to metallocenes. Among these systems, attention have been devoted to complexes containing chelating dialkoxide (*10*), diamido (*11*) and amidinate (*12-15*) ancillary ligations.

Group 4 chelating benzamidinate complexes are normally obtained as a mixture of racemic C_2-symmetry *cis*-octahedral structures. These complexes, when activated with methylalumoxane (MAO), catalyzed the polymerization of ethylene (*16-18*), propylene (*16,19*), styrene (*17*), the oligomerization of 1,5-hexadiene (*16*) and the isomerization of olefins (*20*). Currently, complexes with acetylacetonate (*21*) and phosphinoamide (*22*) ligations, were also found to be active in the polymerization of propylene.

Here we report the synthesis of some group 4 octahedral benzamidinate, acetylacetonate, ketoimidinate and phosphinoamide complexes and their application for the polymerization of propylene, modulated by monomer concentration in the reaction mixture. This is a novel example of modulation of the stereoregularity of polymers by pressure, catalyzed by early transition metal octahedral complexes. The effects of pressure, temperature, nature of solvent in the polymerization process as well as the properties of the obtained polymers, are presented. A strategy for preparing a new type of an elastomeric polypropylene and the mechanism of its formation is also discussed.

Results and Discussions

Synthesis of the Complexes and their Structural Features

The lithium benzamidinate $Li(TMEDA)[p\text{-}Me\text{-}C_6H_4C(NSiMe_3)_2]$ was prepared through the reaction of *p*-toluonitrile and the lithium salt of hexamethyldisilazane. Reactions of MCl_4 with two equivalents of the ligand afforded the dichloride benzamidinates (M = Zr, Ti) as crystalline solids.

$$(1)$$

Alkylation of both compounds with two equivalents of MeLi·LiBr yields the corresponding dimethyl complexes of Zr (**1**) and Ti (**2**) (equation 1).

X-ray studies of **1** (*19*) showed that the central Zr atom is octahedral surrounded by the two benzamidinate ligands and two terminal methyl groups (Figure 1). One C atom from the CH_3 group (C(1)) and nitrogen atom of a chelate unit are in the axial positions, while the C(2) atom and the remaining nitrogen atoms occupy the equatorial positions, producing a C_2-symmetry complex. The two four-membering rings $ZrCN_2$ are almost symmetric and perpendicular to each other. X-ray of the complex **2** shows that it is isostructural with the complex **1**.

Reaction of acetylacetone with BuLi afforded the lithium acetylacetonate ligand. Interaction of two equivalents of the ligand with MCl_4 produced orange powder of the titanium complex (**3**) and light-yellow microcrystals of the zirconium complex (**4**) (equation 2).

$$(2)$$

In the absence of any X-ray data, stereoisomeric studies of acetylacetonate complexes have been conducted using NMR, vibration spectroscopy and dipole

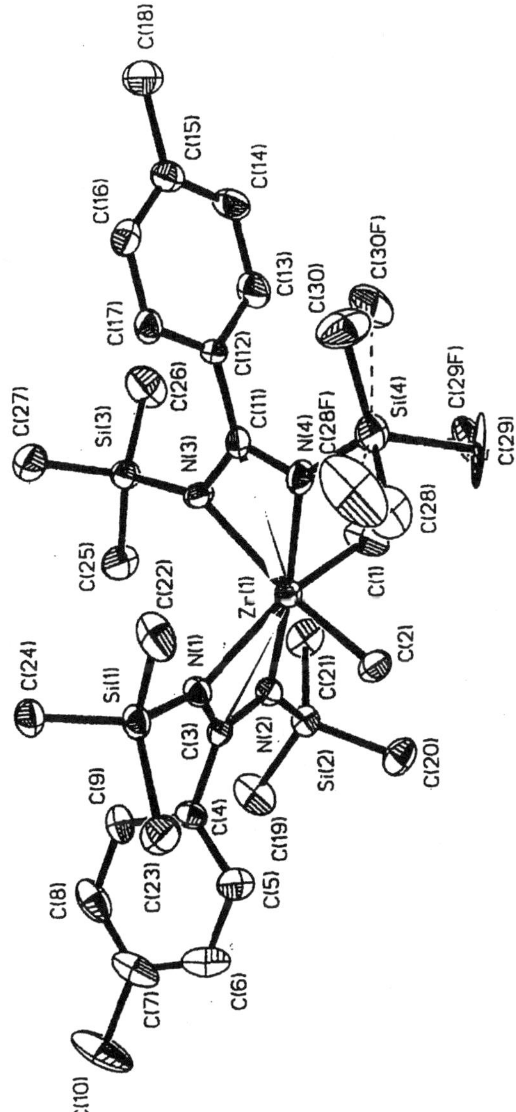

Figure 1. X-ray of complex 1.

moment measurements. These complexes were characterized to be obtained with a *cis*-octahedral configuration (*25-29*).

The Li salt of β-diketoimidinate ligand was obtained by the reaction between C_6H_5CN and $LiCH(SiMe_3)_2$ (*30*). Two equivalents of the ligand react with MCl_4 producing reddish-brown powder of the Ti complex (**5**) and yellow crystals of the Zr complex (**6**) (equation 3).

The Ti complex **5** has an octahedral structure, whereas in complex **6** one of the β-diketiminate ligands undergoes a retrobrook rearrangement to the amidinate isomer catalyzed by $ZrCl_4$ (*30*).

The homoleptic phosphinoamide complex $[Zr(NPhPPh_2)_4]$ (**7**) was prepared by reacting $ZrCl_4$ and four equivalents of $LiNPhPPh_2$ (equation 4), whereas the bisamido complex $[TiCl_2\{N(PPh_2)_2\}_2]$ (**8**) was obtained from $TiCl_4$ and two equivalents of $LiN(PPh_2)_2$ (equation 5). $LiNPhPPh_2$ and $LiN(PPh_2)_2$ were prepared by deprotonation of the correspondent phosphinoamines $Ph_2PN(H)Ph$ (*23*) and $Ph_2PN(H)PPh_2$ (*24*) with butyllithium in hexane.

At room temperature the ^{31}P-NMR spectrum of the titanium complex **8** shows a major signal (90%) at 47.0 ppm which correspondent to the complex with tetrahedral symmetry, and a minor set (10%) of two small doublets at 5.0 and − 10.0 ppm corresponding to the complex with a C_2-symmetry octahedral configuration. The 2D heterocosy spectrum shows that the two doublets are correlated to each other indicating that only one isomer was obtained among all the conceivable complexes that can be theretically formed. The complexes responsible for the signals at 47 and 5.0-10.0 ppm were found to be in equilibrium, as shown by the change in intensity of the signals by changing the temperature. The process is fully reversible by either cooling or warming the sample to room temperature.

Propylene Polymerization Experiments

The catalytic polymerization of propylene was studied using complexes **1-8** and MAO as a cocatalyst. Reactions at atmospheric pressure were carried out in a 100 ml flamed round-bottom flask. Experiments at higher pressures were performed either in 100 ml heavy-wall glass or in stainless steal reactors, equipped with a magnetic stirrer. Reactors were charged inside a glovebox with a catalyst and MAO and connected to a high vacuum line. After introducing certain amount of solvent (toluene or dichloromethane) the reactor was frozen at liquid nitrogen and pumped-down. For the reaction at atmospheric pressure, propylene was used to back filled the flask and maintained at 1.0 atm with a mercury manometer. In experiments at higher pressure propylene was vacuum transferred to the frozen reactor. After stirring for a certain period, the unreacted propylene was vented out and reaction was quenched by addition of HCl/CH_3OH. Polymer was cleaned from the catalytic system residues by dissolving in a hot (120°C) 1,2,4-trichlorobenzene or decaline. After filtration of

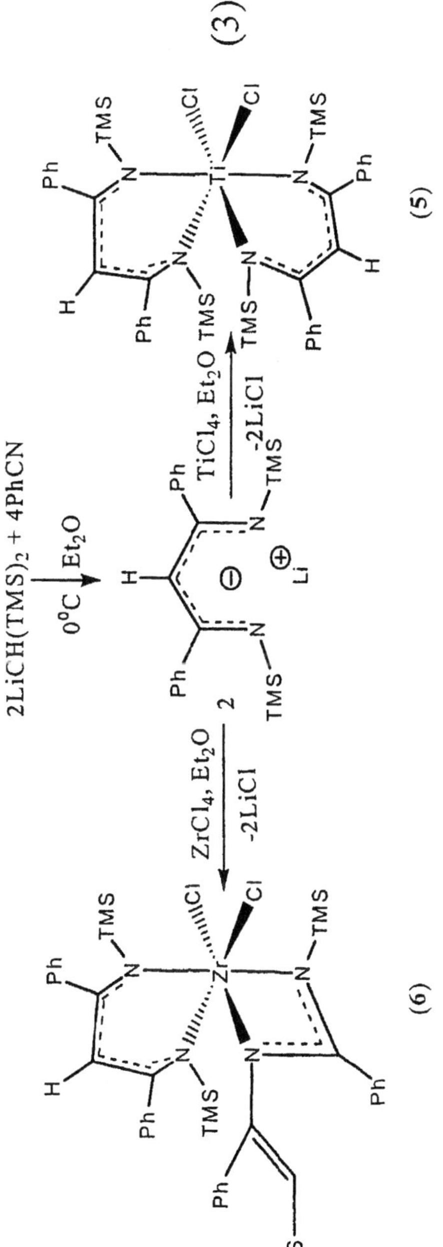
2LiCH(TMS)$_2$ + 4PhCN
0°C, Et$_2$O
Ph
H
Ph
N—TMS
N—TMS
Li
2
TiCl$_4$, Et$_2$O
-2LiCl
ZrCl$_4$, Et$_2$O
-2LiCl
Ph
Ph
TMS
N
N
TMS
Cl
Cl
Ti
N
N
TMS
TMS
Ph
Ph
H
(5)
Ph
Ph
TMS
N
N
TMS
Cl
Cl
Zr
N
N
TMS
TMS
Ph
Ph
H
Ph
S
(6)
(3)

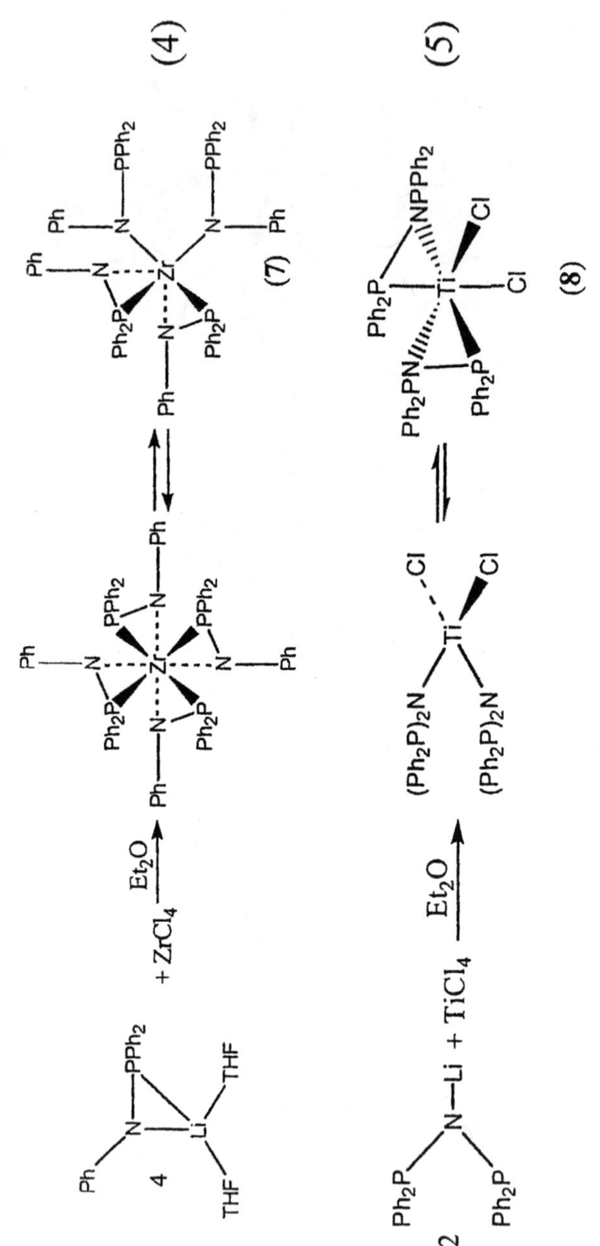

the hot solution, the filtrate was poured into cold acetone to precipitate the polypropylene. The latter was then washed with acetone for removing traces of solvents and dried.

Effect of the Complexes Structure on the Activity and Tacticity in the Polymerization of Propylene.

In general, the tacticity of the polymers predictably differs with the structure of the catalysts. Complexes with C_2 or C_1 symmetry are expected to produce isotactic polyolefins, while compounds with C_{2v} symmetry produce atactic polymers (*3, 31-35*). However, when the polymerization of propylene with complex **1** is performed at atmospheric pressure, unexpectedly instead of an isotactic polymer, an oily atactic product is formed (Table I). At the same time, when the reaction is carried out at high monomer concentration (in liquid propylene), the expected highly stereoregular polypropylene (*mmmm* > 99.5%, m.p. = 153°C) was obtained. This phenomena was rationalized by an intramolecular epimerization reaction of the growing polymer chain (P) at the last-inserted monomeric unit, (Scheme 1) (*36-38*). Apparently, a high propylene concentration allows faster insertion of the monomer and almost completely suppresses the epimerization reaction.

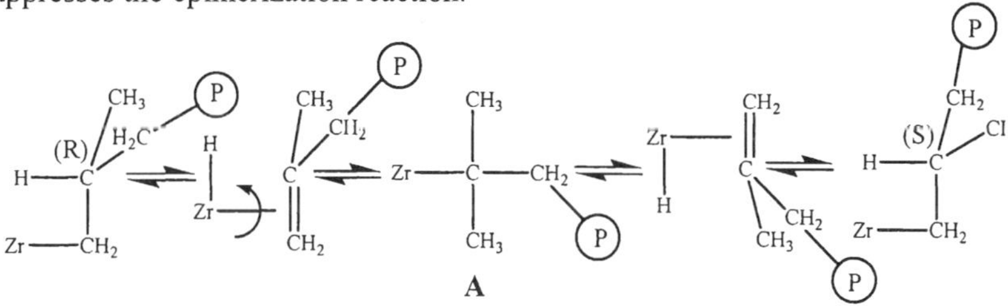

(P) = polymer

Scheme 1. Intramolecular mechanism for the epimerization of polypropylene.

Thus, the stereoregularity of polypropylene with the system " **1**+ MAO" can be generalized as described in equation 6.

An additional corroboration that the mechanism shown in Scheme 1 is responsible for the stereodefects of polypropylene was obtained by transformations of 1-octene in the presence of **1** and MAO (*19-21*). The polymerization of 1-octene in these conditions is extremely low. Hence, a β-hydrogen elimination can take place either from each of the CH_3 groups at the α-position in the intermediate **A** (Scheme 1), or from the CH_2 group of the 1-octene molecule attached to the complex. In the first case no changes in the alkene are expected, but in the second mechanism, an isomerization of the double bond will be induced. It was shown that 1-octene almost completely

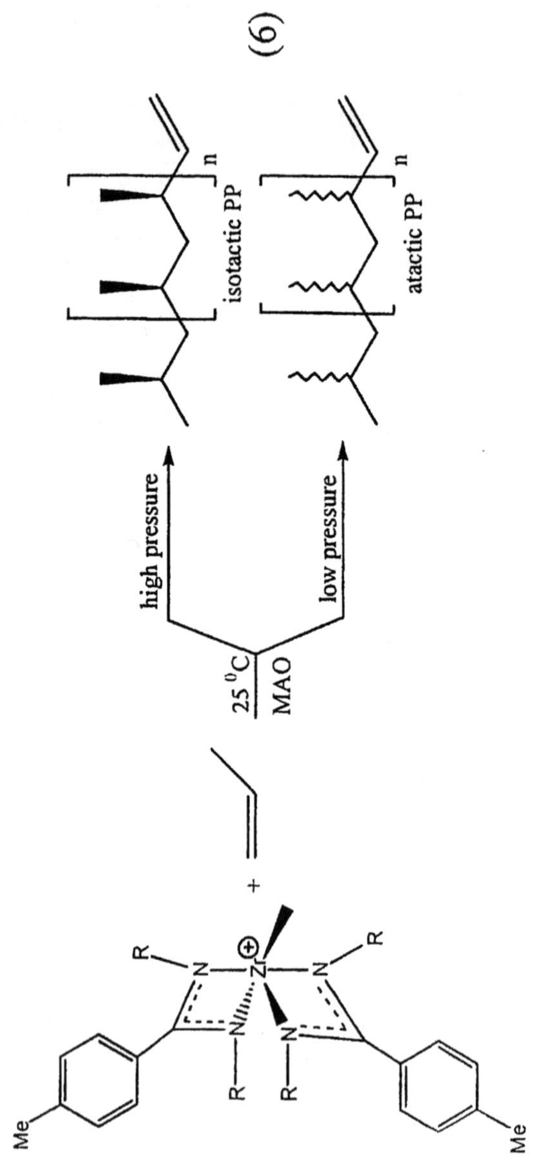
Me
R
R
R
R
25 °C
MAO
high pressure
low pressure
isotactic PP
atactic PP
n
n
Me
(6)

Table I. Data for the Polymerization of Propylene Catalyzed by Benzamidinate Complexes 1 – 8 Activated by Methylalumoxane[a]

Run	C^b	S^c	$Al\!:\!M$	P^d (atm)	A^e	Mn^f	mwd^g	$mmmm$ (%)	$Type^h$
1	1	T	250	1.0	8.1				oil
2	1	T	250	9.2	1.1^i	261	1.69	86^j	i
						36	2.35	11	oil
3	1	D	250	9.2	2.2	11	2.49	90	i
4	1	D	1000	9.2	7.9	58	1.42	98	i
5	2	T	1000	1.0	-	-	-	-	-
6	2	D	1000	1.0	-	-	-	-	-
7	2	T	1000	9.2	1.9	51	1.83	21	e
8	2	D	1000	9.2	1.9	55	1.51	18	e
9	3	T	430	1.0	-	-	-	-	-
10	3	T	430	9.2	1.9	129	1.92	15	e
11	3	D	400	9.2	7.3	35	3.32	22	e
12	4	T	480	1.0	-	-	-	-	-
13	4	T	480	9.2	0.9	53	1.24	99	i
14	4	D	480	9.2	0.2	71	1.82	28	e
15	5	T	400	9.2	1.1	596	2.14	27	e
16	5	D	800	9.2	0.2	-	-	-	oil
17	7	T	500	9.2	0.010	48	2.40		e
18	8	T	500	9.2	0.016	88	1.65		e

[a] – 25°C; [b] – catalyst; [c] - solvent (T = toluene, D = dichloromethane), [d] – pressure, [e] - A = activity x 10^{-5}, g PP/mol cat·h; [f] – Mn x 10^{-3} gr/mol, [g] - molecular weight distribution; [h] – i = isotactic polymer, e = elastomeric polymer, [I] - mixture of isotactic (50%) and atactic (50%) fractions; [j] – isotacticity of isotactic fraction.

isomerized to a mixture of octenes, strongly supporting the formation of intermediate **A**.

As compared to toluene, larger activities and stereoregularities of the polymers are achieved when the polymerization is carried out in CH_2Cl_2 (entries 2 and 3). Plausibly, this polar solvent causes charge separation between the cationic complex and the MAO counter-anion, thus encouraging the insertion of the monomer. For toluene, insertion of the monomer as well as termination of the polymer chain is inhibited by bounding of the ring to the cationic center (*39-41*), resulting in a decrease of the polymerization rate and an increase of polymer's molecular weight.

In contrary to **1**, complex **2** is inactive in the polymerization at atmospheric pressure in both solvents (entries 5 and 6). The solvent-metal interaction seems to be stronger in Ti than in Zr because of the effective charge at the metal center, resulting in an inhibition of the monomer insertion for complex **2**. At higher pressure, complex **2** is an active precatalyst in both solvents producing an elastomeric polymer.

Thus, group 4 *bis*(benzamidinate) complexes activated by MAO can be considered as a new catalytic system for the polymerization of propylene, when the stereoregularity of the polymers can be modulated by pressure (from atactic to isotactic through elastomers) (*19*). This fact raises a conceptual question regards whether simple octahedral or even tetrahedral complexes having a dynamic Lewis-base pendant group donating an electron pair to the electrophilic metal center, are suitable for the synthesis of a unique product – an elastomeric polypropylene. As is shown in Scheme 2, a dynamic equilibrium occurs between tetrahedral and *cis*-octahedral configurations (X = halide, E = donor group with a electron lone pair, R = C, N, P or other anionic bridging group). A plausible *trans*-octahedral complex which can be formed in this type of dynamic process, is unable to perform the olefin insertion and has no catalytic activity (*3,11, 31,32*).

Results of the polymerization of propylene with complexes **3-8** activated by MAO are also shown in Table 1. Propylene does not polymerizes with complex **3** at atmospheric pressure (entry 9) whetreas at higher pressures an elastomeric polymer is formed. As well as in polymerization with the benzamidinate complex **2**, an increase in activity is observed when CH_2Cl_2 was used, as compared with toluene. At the same time, the polymers obtained in toluene have larger molecular weights (entries 10 and 11). Similarly as with complex **3**, in the reaction at atmospheric pressure with complex **4** polypropylene was not produced. In both solvents at higher pressure, complex **3** was found to be more active than **4**, whereas the latter produced more stereoregular polymers (entries 10,11 and 13,14). In toluene, complex **4** yields two polymeric fractions: 76% of a highly isotactic polymer (m.p.= 156°C) and 24% of an elastomeric polymer. This fact argues that in toluene, the cationic form of complex **4** may be in two isomeric forms, responsible for two different polymerization pathways.

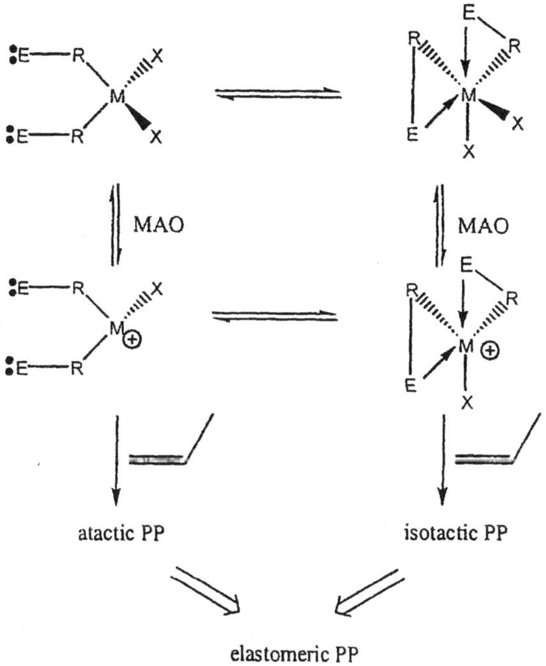

Scheme 2. Elastomeric polypropylene obtained by the dynamic interconversion between cationic complexes with C_2v and C_2 symmetry.

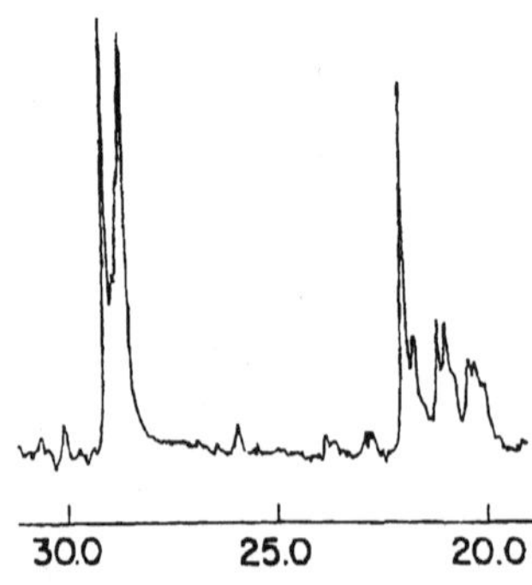

$$6 \ + \ \text{MAO} \ \longrightarrow \tag{7}$$

When complex **5** is activated by MAO in toluene, it catalyzes the formation of an elastomeric polypropylene (entry 15). Contrary to **5**, the Zr complex **6** showed a negligible activity that may be a consequence of an interaction of **6** with MAO (equation 7). The double bond in the ligand, being in a close proximity, probably occupies the free coordinative site, more likely than a monomer molecule, thus rendering the complex almost inactive for the polymerization reaction.

Regarding the dynamic complexes, the activity of the titanium complex **8** is much higher than that of complex **7**. At the reaction conditions no high molecular weight atactic polymer was formed, but only an elastomeric polymer, in accordance with the postulated equilibrium between a tetrahedral and a *cis*-octahedral active cationic species (Scheme 2). The elastomeric products obtained with **7** and **8**, are characterized by narrow polydispersities suggesting on a single-site character of these precatalysts.

Structure and Properties of the Elastomeric Polypropylenes

The ^{13}C-NMR spectrum of the elastomeric polymers obtained with complex **8** (Figure 2) exhibits in the methyl region a large signals of the *mmmm* pentad, ruling out the production of a high molecular weight atactic polymers.

Figure 2. ^{13}C{H} NMR methyl region of the elastomeric polypropylene obtained with complex 8.

The elastomeric polypropylene obtained with complex **3** has a Tg temperature of -6.7°C and exhibits a broad melting region from 137 to 156°C. Mechanical properties of the polymer indicate a tension strength of $\sigma = 157$ kg/cm^2 when extended by 160%. The polymer shows a remarkable memory effect, returning to its initial position.

The ^{13}C-NMR spectra of the elastomeric polymer obtained with **2** was compared with the spectrum of "atactic oil" (*mmmm* = 7%) and "isotactic polypropylene" (*mmmm* ~ 90%) prepared with **1**. Based on the NMR data, the statistical lengths of the isotactic blocks between two neighboring epimerization stereodefects were calculated. For the elastomeric polymer, the length of the isotactic fragment between two stereodefects was found to be shorter (11-27 CH$_3$ groups) than that obtained for the isotactic polymer (35-45 CH$_3$ groups), and longer than for the atactic sample (7-8 CH$_3$ groups). Thus, in contrary to the high molecular weight atactic polymer (*42*), in our samples long isotactic domains are disposed between stereodefects. On the other hand, differing to the elastomers where the alternating isotactic and atactic blocks have commensurate lengths (*43-46*), our samples are characterized by frequent alternation of the isotactic domains with many stereodefects. As compared to the elastomers prepared through the "Dual-Side" mechanism (*47*), which exhibit *mmmm* content up to 72%, in our case the polymers resides with lower stereoregularity. In the "Dual-Side" mechanism, each site of the catalyst differs by its symmetry, and with the back skipping of the polymeric chain, two stereopossibilities for the insertion of the monomer molecule are obtained. In our complexes, a back skipping of the chain will induce the same symmetry, thus providing the same type of stereoregular insertion. The stereoregular errors in the polymer are formed because of the intramolecular epimerization of the growing chain at the last inserted unit. The rate correlation of the epimerization with the insertion influences the amount of the isotactic domains between the stereodefects.

Conclusion

The presented data indicate that octahedral cationic complexes with C$_2$-symmetry activated with methylalumoxane catalyze the polymerization of propylene towards isotactic polypropylene, whereas tetrahedral complexes with C$_{2v}$-symmetry lead to formation of an atactic polymer. The rapid interconversion of these two structures results in formation of elastomeric polypropylene. The results obtained allow to shed light concerning the applicability of the *cis*-octahedral C$_2$-symmetry metalloorganic complexes for the stereospecific polymerization of α-olefins. The polymers stereoregularity can be modulated by pressure, allowing the formation of highly isotactic or elastomeric polymers. Varying the solvent and metal coordination can also control the stereoregularity of the polymers. It seems that the crucial factor in the polymerization is the symmetry of the catalyst, regardless to the structure of the ancillary ligands. The large variety of synthesized ligands is an attractive incentive for the development of new octahedral catalysts for the preparation of new types of polymeric materials.

References

1. Kaminsky, W. Olefin Polymerization Catalyzed by Metallocenes. *In Advances in Catalysis*; Gates B. C.; Knözinger, H., Eds. Academic Press: San Diego, 2002; Vol. 46.

2. Mülhaupt, R. in *Polypropylene: An A-Z reference*; Karger-Kocsis, J., Ed.; Kluwer Academic Publisher: Dordrecht, 1999, 454-475.

3. Kaminsky, W.; Arndt, M. *Adv. Polym. Sci.* **1997**, *127*, 143-187.

4. Alt. H. G.; Köpll, A. *Chem. Rev.* **2000**, *100*, 1205-1222.

5. Coates. G. W. *Chem.Rev.* **2000**, *100*, 1223-1252.

6. Chen, Y.-X.; Stern, C. L.; Yang, S.; Marks, T. J. *J. Am. Chem. Soc.* **1996**, *118*, 12451-12452.

7. Galan-Fereres, M.; Koch, T.; Hey-Hawkins, E.; Eisen, M. S. *J. Organomet. Chem.* **1999**, *580*, 145-155.

8. Ittel, S. D.; Johnson, L. K.; Brookhart, M. *Chem. Rev.* **2000**, *100*, 1169-1204.

9. Britovsek, G. J. P.; Gibson, V. C.; Wass, D. F. *Angev. Chem., Int. Ed. Engl.* **1999**, *38*, 428-447.

10. Linden, A. V. D.; Schaverien, C. J.; Meijboom, N.; Ganter, C.; Orpen, A. J. *J. Am. Chem. Soc.* **1995**, *117*, 3008-3021.

11. Mack, H.; Eisen, M. S. *J. Organomet. Chem.* **1996**, *525*, 81-87.

12. Richter, J.; Edelmann, F. T.; Noltemeyer, M.; Schmidt, H.-G.; Shmulinson, M.; Eisen, M. S. *J. Mol. Catal.* **1998**, *130*, 149-162.

13. Jayaratne, K. C.; Sita, L. R. *J. Am. Chem. Soc.* **2001**, *123*, 10754-10755.

14. Hagadorn J. R.; Arnold, J. *Organometallics* **1998**, *17*, 1355-1368.

15. Edelmann, F. T. *Coord. Chem. Rev.* **1994**, *137*, 403-481.

16. Walter, D.; Fischer, R.; Görls, H.; Koch, J.; Scheweder, B. *J. Organomet. Chem.* **1996**, *508*, 13-22.

17. Flores, J. C.; Chien, J. C. W.; Rausch, M. D. *Organometallics* **1995**, *14*, 2106-2108.

18. Hershcovics-Korine, D.; Eisen, M. S. *J. Organomet. Chem.* **1995**, *503*, 307-314.

19. Volkis, V.; Shmulinson, M.; Averbuj, C.; Lisovskii, A.; Edelmann, F. T.; Eisen, M. S. *Organometallics* **1998**, *17*, 3155-3157.

20. Averbuj, C.; Eisen, M. S. *J. Am. Chem. Soc.* **1999**, *121*, 8755-8759.

21. Shmulinson, M.; Galan-Fereres, M.; Lisovskii, A.; Nelkenbaum, E.; Semiat, R.; Eisen, M. S. *Organometallics* **2000**, *19*, 1208-1210.

22. Kühl, O.; Koch, T.; Somoza Jr, F. B.; Junk, P. C.; Hey-Hawkins, E.; Plat, D.; Eisen, M. S. *J. Organomet. Chem.* **2000**, *604*, 116-125.

23. Ashby, M. T.; Li, Z. *Inorg. Chem.* **1992**, *31*, 1321-1322.

24. Schmidbaur, H.; Lauteschläger, S.; Köhler, F. H. *J. Organomet. Chem.* **1984**, *271*, 173-180.

25. Pinnavaia, T. J.; Fay, R. C. *Inorg. Chem.* **1968**, *7*, 502-508.

26. Serpone, N.; Fay, R. C. *Inorg. Chem.* **1969**, *8*, 2379-2384.

27. Fay, R. C.; Lowry, R. N. *Inorg. Chem.* **1967**, *6*, 1512-1519.

28. Comba, P.; Jakob, H.; Nuber, B.; Keppler, B. K. *Inorg. Chem.* **1994**, *33*, 3396-3400.

29. Morstein, M. *Inorg. Chem.* **1999**, *38*, 125-131.

30. Hitchcock, P. B.; Lappert, M. F.; Liu, D. -S. *J. Organomet. Chem.* **1995**, *488*, 241-248.

31. Bochmann, M. *J. Chem. Soc. Dalton Trans.* **1996**, 225-270.

32. Brinzinger, H. H.; Fisher, D.; Mülhaupt, R.; Rieger, B.; Waymouth, R. M. *Agnew. Chem., Int. Ed. Engl.* **1995**, *34*, 1143-1170.

33. Averbuj, C.; Tish, E.; Eisen, M. S. *J. Am. Chem. Soc.* **1998**, *120*, 8640-8646.

34. Ewen, J. A.; Jones, R. L.; Razavi, A.; Ferrara, J. D. *J. Am. Chem. Soc.* **1998**, *110*, 6255-6256.

35. Möhring, P. C.; Coville, N. J. *J. Organomet. Chem.* **1994**, *479*, 1-29.

36. Busico, V.; Cipillo, R.; Caporaso, L.; Angeleni, G.; Segre, A. L. *J. Mol. Catal. Part A.* **1998**, *128*, 53-64.

37. Busico, V.; Brita, D.; Caporaso, L.; Cipillo, R.; Vacatello, M. *Macromolecules* **1997**, *30*, 3971-3977.

38. Busico, V.; Caporaso, L.; Cipillo, L.; Landriani, L.; Angeleni, G.; Margonelli, A.; Segre, A. L. *J. Am. Chem. Soc.* **1996**, *118*, 2105-2106.

39. Eisen, M. S.; Marks, T. J. *J. Am. Chem. Soc.* **1992**, *114*, 10358-10368.

40. Eisen, M. S.; Marks, T. J. *Organometallics* **1992**, *11*, 3939-3941.

41. Lancaster, S. J.; Robinson, O. B.; Bochmann, M.; Coles, S. J.; Hursthouse, M. B. *Organometallics* **1995**, *14*, 2456-2462.

42. Resconi, L.; Jones, R. L.; Rheingold, A. L.; Yap, G. P. A. *Organometallics* **1996**, *15*, 998-1005.

43. Tagge, C. D.; Kravchenko, R. L.; Lal, T. K.; Waymouth, R. M. *Organometallics* **1999**, *18*, 380-388.

44. Maciejewski, P.; Agoston, T.; Lal, T. K.; Waymouth, R. M. *J. Am. Chem. Soc.* **1998**, *120*, 11316-11322.

45. Hu, Y.; Krejchi, M. T.; Shah, C. D.; Myers, C. L.; Waymouth, R. M. *Macromolecules* **1998**, *31*, 6908-6916.

46. Bruce, M. D.; Coates, G. W.; Hauptman, E.; Waymouth, R. M.; Ziller, J. W. *J. Am. Chem. Soc.* **1997**, *119*, 1117-11182.

47. Dietrich, U.; Hackmann, M.; Rieger, B.; Klinga, M.; Leskela, M. *J. Am. Chem. Soc.* **1999**, *121*, 4348-4355.

Chapter 5

Complexes of Amine Phenolate Ligands as Catalysts for Polymerization of α-Olefins

Moshe Kol[1], Edit Y. Tshuva[1], and Zeev Goldschmidt[2]

[1]School of Chemistry, Tel Aviv University, Ramat Aviv,
69978 Tel Aviv, Israel
[2]Department of Chemistry, Bar-Ilan University, Ramat-Gan 52900, Israel

The coordination chemistry and the catalytic potential in polymerization of alpha-olefins of two families of group IV metal complexes of amine phenolate ligands are outlined. The first family is based on dianionic amine diphenolate ligands. These ligands lead to C_s-symmetrical complexes, in which the two labile groups are in cis geometry. An additional donor on the sidearm was found to be crucial for the activity of the resulting catalysts. Zirconium and hafnium based catalysts led to extremely high activities in polymerization of 1-hexene, whereas titanium catalysts led to living polymerization at room temperature. The second family is based on diamine diphenolate ligands that lead to C_2-symmetrical complexes that may afford isotactic and living polymerization at room temperature, depending on the bulk of the phenolate substituents.

In recent years there has been a broad interest in developing of cyclopentadienyl-free catalysts based on group IV metals for polymerization of alpha-olefins (*1*). Some recent milestones include the living polymerization of high olefins by diamido based systems introduced by McConville (*2*) and Schrock (*3*); the extremely active phenoxy-imine catalysts for polymerization of ethylene introduced by Fujita (*4*); and the living and syndiotactic polymerization of propylene by the phenoxy-imine systems introduced by Coates (*5*), and Fujita (*6*). These studies show that by careful design of the ligand system, a close control of the activity of the catalyst and the properties of the polymer may be achieved. Recently, we introduced two families of catalysts for polymerization of alpha-olefins based on group IV metal complexes of chelating dianionic amine phenolate ligands. In this chapter we outline the main attributes of these catalytic systems.

Amine Diphenolate Based Catalysts

Design and Synthesis of the Ligands

The ligand precursors employed in this section are bis(2-hydroxyarylmethyl)amines. A large variety of these compounds are known, and several reports that described their application as precursors for $[ONO]^{2-}$ ligands for late transition metals have appeared (*7, 8*), however, prior to our work (*9*), they have not been used as ligands for Group IV transition metals. Upon binding to a metal, a stable geometry featuring two rigid six-membered metalla-rings is expected to form, as shown in Figure 1. The two remaining labile groups render this complex suitable to serve as a precatalyst for polymerization of alpha-olefins.

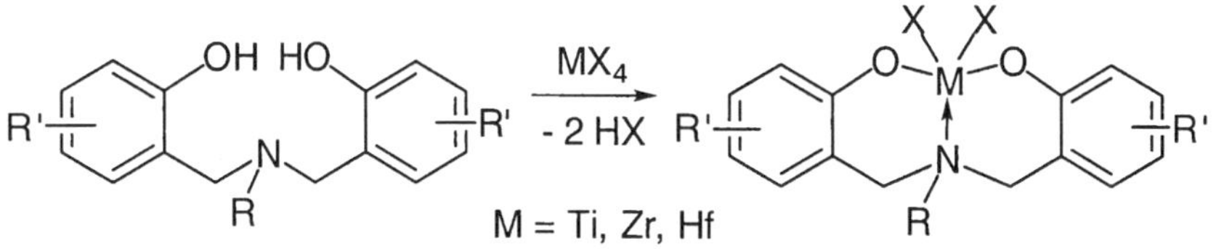

Figure 1. General structure of the ligand precursor and its group IV metal complexes.

The amine diphenolate ligand precursors are easily synthesized by a single-step Mannich-type condensation between a primary amine, two equiv of formaldehyde, and two equiv of a substituted phenol (*10*), as shown in Figure 2. As a large variety of both primary amines and substituted phenols are available, a fine tuning of the resulting ligand may be achieved enabling a close control of the steric and electronic properties of the resulting metal complex.

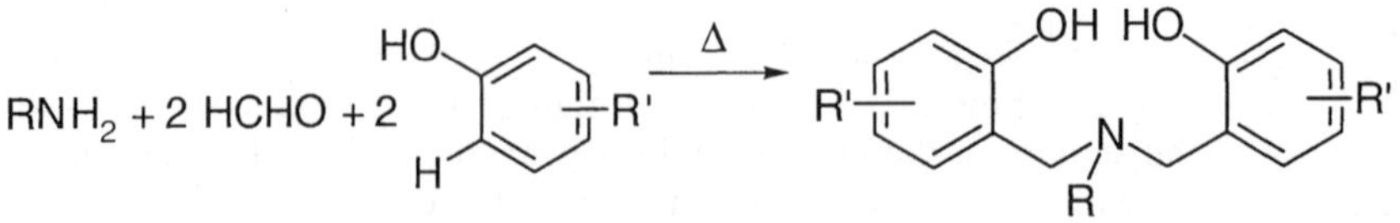

Figure 2. Synthesis of the amine diphenolate ligand precursors.

Our preliminary studies indicated that the strongest effect of a ligand structural motif on the reactivity of the resulting complexes was the presence of an extra donor on the sidearm bound to the central nitrogen donor (*11*). Ligands lacking this donor are termed [ONO]-type ligands, whereas lignds having such a donor are termed [ONXO]-type ligands. The nature of the X-donor (nitrogen, oxygen and sulfur), its hybridization (sp^3 or sp^2), its bulk, and the distance between it and the central nitrogen donor were all found to affect the reactivity of the resulting catalyst. The steric bulk of the phenolate groups (e.g. *tert*-butyl vs. methyl substituents in the ortho position to the phenolate oxygen) was found to have a more subtle effect on reactivity. A representative selection of these ligands is shown in Figure 3.

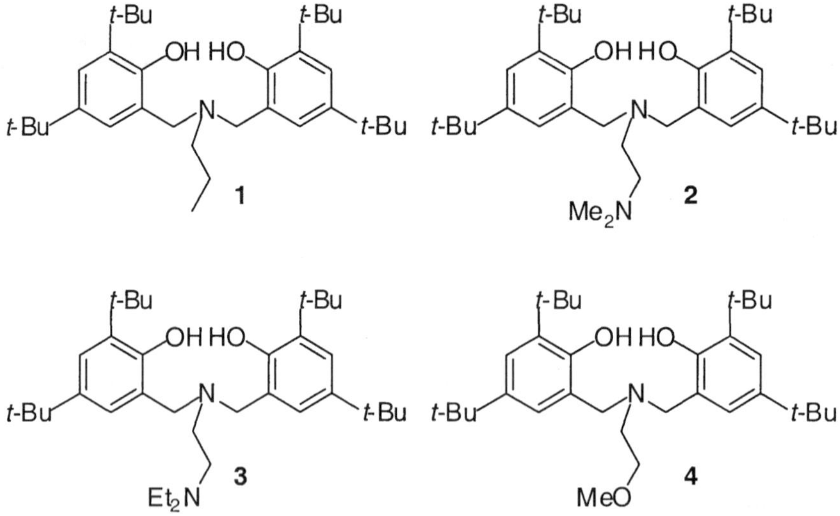

Figure 3. Representative amine diphenolate ligand precursors: 1 – an [ONO]-type ligand precursor; 2-4 – [ONXO]-type ligand precursors.

Zirconium and Hafnium Complexes – Synthesis, Structure and Reactivity

The most straightforward entry to zirconium and hafnium dialkyl complexes is the single-step metathesis reaction between the ligand precursors and the corresponding tetrabenzyl metal compounds (*11–13*). These reactions take place between RT and 65 °C. The presence of the bulky ortho *tert*-butyl substituent on the phenolate rings induces very clean transformations, giving the desired products in essentially quantitative yields. Ligand precursors having smaller substituents lead to lower yields, probably due to formation of bis(homoleptic) complexes (*12, 14*).

The pentacoordinate [ONO]MBn$_2$ complexes (M = Zr, Hf) are expected to be of C_s-symmetry, whereas the hexaccordinate [ONXO]MBn$_2$ complexes are expected to be of either C_s-symmetry or C_1-symmetry, as shown in Figure 4. For both symmetries of the hexaccordinate complexes, a cis relationship between the two labile positions is dictated by the [ONXO]-type ligand, that may facilitate the formation of a metallacyclobutane in the catalytic polymerization cycle.

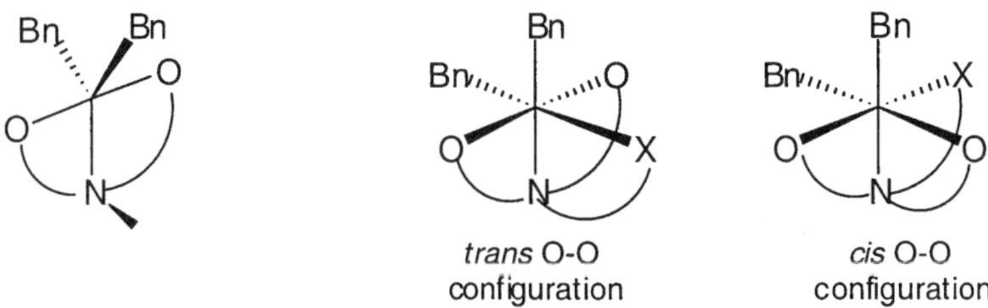

Figure 4. Expected geometries for [ONO]MBn$_2$ and [ONXO]MBn$_2$ complexes.

The ^{1}H NMR spectrum of the [ONO]ZrBn$_2$ complex **5**, synthesized from ligand precursor **1** above, indicated the formation of a single isomer of C_s-symmetry, and its X-ray structure supported this notion. The structure of **5** features (Figure 5, left) a complex of pseudo trigonal bipyramidal geometry, with axial O atoms and equatorial N, C, C atoms. The C-Zr-C angle between the two benzyl groups is 117.4°. An acute Zr-CH$_2$-C(Ar) angle of 89.4° and a short Zr-C(Ar) distance of 2.71 Å for the 'top' benzyl group support a non-classical η^2-binding of the aromatic group to the metal.

The symmetry of the [ONXO]MR$_2$ complexes (R = alkyl, chloro, alkoxo, etc.) was found to depend on the nature the R-groups (*15*). When R is a benzyl group, only the C_s-symmetrical complexes were obtained. The X-ray structure of **6**, a zirconium complex synthesized from ligand precursor **2** above, having a dimethylamino sidearm donor is shown in Figure 5, right.

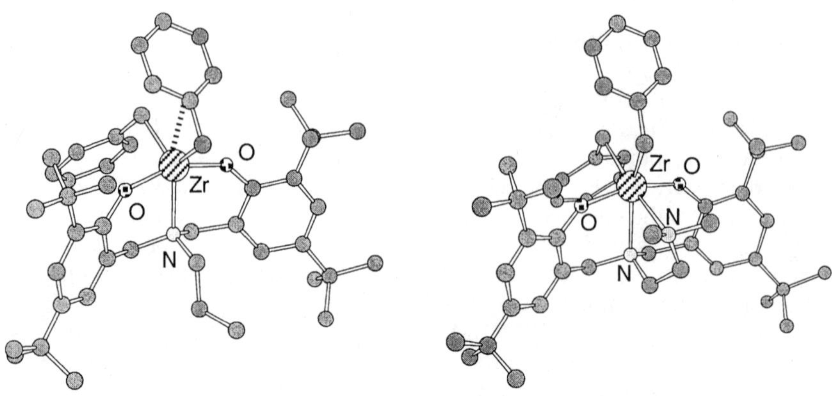

*Figure 5. X-ray structure of the [ONO]ZrBn₂ complex **5** (left) and the [ONXO]ZrBn₂ complex **6** (right).*

6 is a slightly distorted octahedral complex with trans disposition of the O atoms. The sidearm donor is bound to the metal, and the angle between the two labile benzyl groups, both bound in a η^1 fasion, is much narrower in comparison to **5** (93.7°). It is noteworthy that despite the difference in coordination numbers and benzyl hapticities between **5** and **6**, the 'cores' of these two complexes are similar (*12*). This is manifested in similar O-Zr and N-Zr bond lengths, and back bending of the aromatic rings away from the 'equatorial' benzyl group. The X-ray structures of various [ONXO]-type zirconium and hafnium complexes have been solved, and all feature the same cores (*13*).

The reactivity of these two types of dibenzyl complexes was studied by activating them with tris(pentafluorophenyl)borane in neat 1-hexene at room temperature. Unexpectedly, they exhibited a dramatic difference in reactivity. The [ONO]ZrBn₂ complex **5** proved to be a poor polymerization precatalyst yielding only traces of oligomeric materials, signifying a rapid catalyst deactivation. The analogous hafnium complex exhibited the same behaviour. In contrast, the [ONXO]ZrBn₂ complex **6**, proved to be an exceptionally active catalyst for polymerization of 1-hexene. The highest activity obtained thus far in polymerization by **6**, was measured by activating 13 μmol of **6** with 1 equiv of $B(C_6F_5)_3$ in 100 mL of neat 1-hexene. A fast polymerization reaction was initiated accompanied by considerable heat release causing boiling of the monomer, and yielded 22.3 g of poly(1-hexene) after 5 min, corresponding to an activity of 21,000 g mmol cat⁻¹ h⁻¹. The same high activity was observed in polymerization of 1-octene. Under these conditions, the polymer obtained has a relatively low molecular weight of 35,000 and broad molecular weight distribution of $M_w/M_n = 3.5$, due to the high reaction temperature. Taming the polymerization by diluting the monomer in heptane allowed the production of poly(1-hexene) having higher molecular weight (M_w = 170,000) and narrow molecular weight distribution (M_w/M_n = 2.0). The poly(1-hexene) obtained with these catalysts is atactic, as a consequence of the symmetry plane bisecting the two active positions.

These findings indicate that a sidearm donor is essential for obtaining highly active zirconium and hafnium complexes. In order to obtain further insight on the effect of the sidearm donor and attempt to broaden the scope of reactivities, we introduced gradual modifications in the ligands, and studied their effect on the reactivity of the resulting complexes. Several such modifications are outlined below.

The first modification we introduced was changing the bridging unit between the central and the sidearm nitrogens from an ethylene to a propylene bridge. The resulting dibenzyl zirconium complex **7**, was synthesized in quantitative yield and its ^{1}H NMR spectrum supported its C_s-symmetry. Unexpectedly, this complex led only to traces of oligomeric material when activated with $B(C_6F_5)_3$ in neat 1-hexene, signifying a rapid catalyst deactivation. The origin of this effect was revealed by the X-ray structure of that complex, that indicated that the sidearm donor was not bound to the metal, namely, a six-membered chelate is not formed (*12*). Therefore, although this complex is formally an [ONXO]-type complex, it is in fact an [ONO]-type complex. Close inspection of its bond lengths and angles shows that it is isostructural to **5**. These results support the notion that the sidearm donor has to bind to the metal in the reactive species in order to induce the high activity and the high turnover number.

The steric bulk of the sidearm donor was found to play a major role as well. Ligand **3**, featuring a bulkier diethylamino sidearm donor, led to the [ONXO]-type zirconium dibenzyl complex **8**, whose NMR spectra supported its C_s-symmetry. The X-ray structure of **8** indicated that the bond length between the sidearm donor and the zirconium is considerably longer in comparison to **6** (2.76 vs 2.59 Å, respectively). Activating **8** with $B(C_6F_5)_3$ in neat 1-hexene initiated a slow reaction (60 g mmol cat^{-1} h^{-1}), yielding oligomers rather than high molecular weight polymers (M_w = 1,100, PDI = 1.6), according to GPC analysis. This catalyst is active for at least 7 h. The ^{1}H NMR spectrum of the oligomers obtained, includes vinyl signals (4.75, 4.69 ppm) corresponding to the formation of disubstituted terminal olefins. No evidence for internal olefins was found. It is therefore plausible that the termination processes responsible for the low molecular weight result from high statistic termination processes occurring in a regular 1,2-insertion reaction. The difference in activity between **8** and **6** may be assigned to the weak binding of the sidearm donor to the metal, as expressed in the X-ray structure. Two possible binding modes for the sidearm in the reactive catalyst are either a constant weak binding, or an 'on-off' binding, which may explain the unique reactivity pattern of this catalyst: a long lived oligomerization catalyst. This reactivity calls for a relatively high termination / propagation rate ratio, as well as reinitiation of the reactive species.

Zirconium and hafnium complexes of [ONXO]-type ligands featuring either methoxy or methylthio sidearm donors bound via an ethylene bridge to the central nitrogen donor, all led to remarkably active polymerization catalysts for high olefins (*13*). The most active of those was the zirconium complex of ligand **4** featuring a methoxy sidearm donor (an unprecedented activity of 50,000 g mmol cat^{-1} h^{-1}, in neat 1-hexene following activation with B(C$_6$F$_5$)$_3$). Interestingly, the hafnium series exhibited a different order of activities than the zirconium series. The activity order of the zirconium complexes as a function of the sidearm donor was found to be OMe > NMe$_2$ > SMe (5:2:1), whereas the activity order in the hafnium series was found to be SMe > OMe > NMe$_2$ (3.5:1.5:0.2). The X-ray structures of the respective zirconium and hafnium complexes were found to be almost superimposable. The extreme activity of the [ONSO]-hafnium complex may be attributed to the strong interaction between the soft Lewis acid hafnium, and the soft Lewis base sulfur donor.

Characterization of the Reactive species

To gain insight on the structure of the reactive species in solution, we performed various NMR experiments. In particular, NOESY experiments were conducted to determine spatial relationships between different groups in the molecules in the precatalyst and in the reactive species. NOE-correlations between the methyl protons of the dimethylamino donor and the ortho *tert*-butyl group as well as one of the benzyl groups in **6**, supported the hypothesis that the sidearm donor is bound to the metal in solution as well (Figure 6, left).

Reaction of **6** with one equiv of B(C$_6$F$_5$)$_3$ in d$_5$-chlorobenzene resulted in the formation of the activated compound that features a C_s-symmetry similar to that of the precatalyst. ^{19}F NMR implied that a well separated ion pair is obtained (*16*), a finding supported by the ^{1}H NMR chemical shift of one of the benzyl units, which is consistent with a non-coordinated [PhCH$_2$B(C$_6$F$_5$)$_3$]$^-$ anion. This notion was further supported by no NOE-correlation of the abstracted benzyl unit to any protons of the cation. The remaining benzyl unit correlates with both the ortho *tert*-butyl group and the dimethylamino arm, indicating that a mono-cation was formed (Figure 6, right). Moreover, addition of 5 equiv of B(C$_6$F$_5$)$_3$ did not affect the ^{1}H NMR and the ^{19}F NMR of the product, indicating that a dicationic species is not formed under these conditions. The NOE correlation of the remaining benzyl unit with the dimethylamino arm supports two important structural features: (1) the sidearm donor is bound to the metal in the reactive species as well; (2) the position of the remaining benzyl group is similar to that of the 'axial' benzyl in the precatalyst, rather than to that of the 'equatorial' benzyl group.

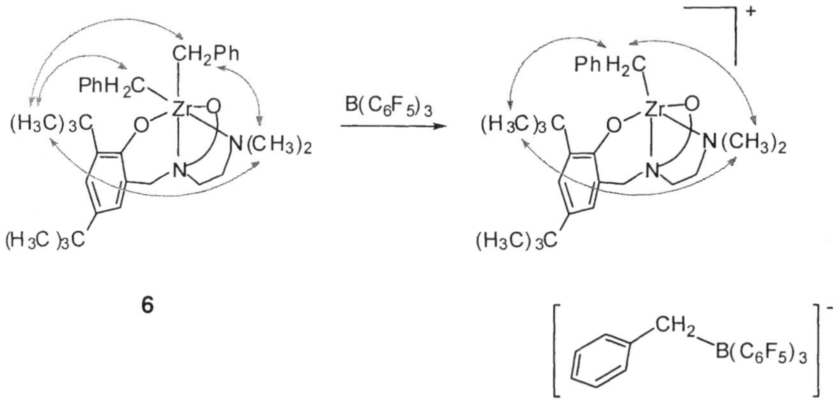

Figure 6. *NOESY correlations in the precatalyst **6** and in the reactive species.*

The same NMR techniques were applied for several other zirconium and hafnium complexes, and all supported the above findings, namely, a well separated ion pair is formed, in which the sidearm donor is bound to the metal and the remaining benzyl group occupies an axial position.

Titanium Complexes: Living Polymerization Catalysts

Titanium catalysts often exhibit different polymerization reactivity than the analogous zirconium or hafnium catalysts. We were therefore interested in developing synthetic routes to amine diphenolate titanium complexes and studying their potential as polymerization catalysts. In particular, we wanted to reveal whether a sidearm donor would have the same effect on reactivity as found for the zirconium and the hafnium complexes.

Two routes for the synthesis of both [ONO]TiBn$_2$ and [ONXO]TiBn$_2$ type complexes were developed. The first route is a three-step one-pot reaction (*17*). The ligand precursors are reacted with titanium tetra(isopropoxide) yielding the corresponding di(isopropoxide) complex. This complex is reacted further with trimethylsilyl chloride, and the resulting dichloro complex is alkylated with two equiv of benzyl magnesium chloride to give the desired complexes in ca. 90% yield. The alternative route is analogous to that used for making the zirconium and hafnium complexes, namely, a direct reaction between the ligand precursors and tetrabenzyl titanium (Figure 7). The instability of the latter compound requires its fresh synthesis.

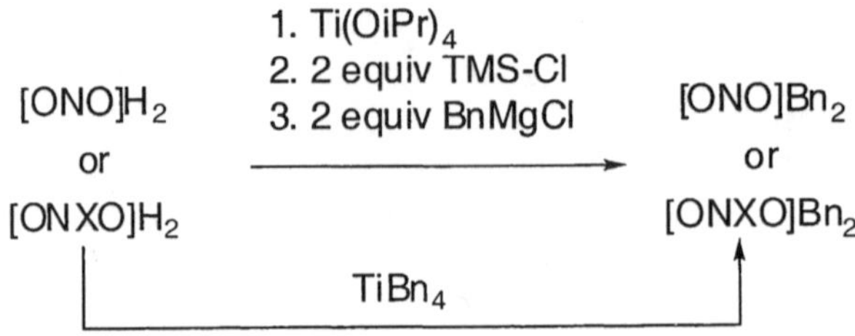

Figure 7. Synthetic routes for dibenzyl titanium complexes.

In contrast to the zirconium and the hafnium series, the lack of sidearm donor in the titanium series did not lead to rapid catalyst deactivation. Thus, activation of the [ONO]TiBn$_2$ complex derived from ligand precursor **1** with B(C$_6$F$_5$)$_3$ in neat 1-hexene at room temperature yielded a mild oligomerization catalyst having an activity of 56 g mmol cat^{-1} h^{-1} (M_n = 1,500; PDI = 2.0).

The presence of a sidearm donor was found to inhibit the termination processes, leading, in certain cases, to living polymerization catalysts at room temperature (*17, 18*). The [ONXO]TiBn$_2$ catalyst exhibiting the most pronounced living properties reported thus far, is that having a methoxy sidearm donor (derived from ligand precursor **4**). The activity of that complex was not high (20 – 35 g mmol cat^{-1} h^{-1}), however it retained its living nature for an exceptionally long time of 31 hours, leading to poly(1-hexene) of M_w = 445,000 and PDI of 1.12. This catalyst was also employed in the block co-polymerization of 1-hexene and 1-octene at room temperature. Thus, 1.5 hours after the full consumption of 110 equiv of 1-hexene by this catalyst (with chlorobenzene as the solvent), 1-octene was added, causing the resumption of the polymerization process. Stopping the polymrization after 3 hours, yielded poly(1-hexene-block-1-octene) having Mn = 11,600 and PDI of 1.2.

Further studies indicated that catalysts retaining their living nature for an even longer period and leading to polymers of higher molecular weights and narrower molecular weight distributions can be developed (*15*).

Diamine Diphenolate Based Catalysts

Design and Synthesis of Ligands

All the complexes of the first family described above, feature a C_s-symmetry. Thus, the olefin may approach each active position from the two possible directions without preference, leading to an atactic polymer. To try and induce tacticity, we aimed at complexes of a different symmetry, that incorporate ligands having similar functional groups yet having a different connectivity mode. Our approach was based on replacing the "branched" connectivity mode of donor atoms with a sequential connectivity mode, namely diamine diphenolate ligands (*19*). Representative ligands of each family, that are constitutional isomers, are shown in Figure 8.

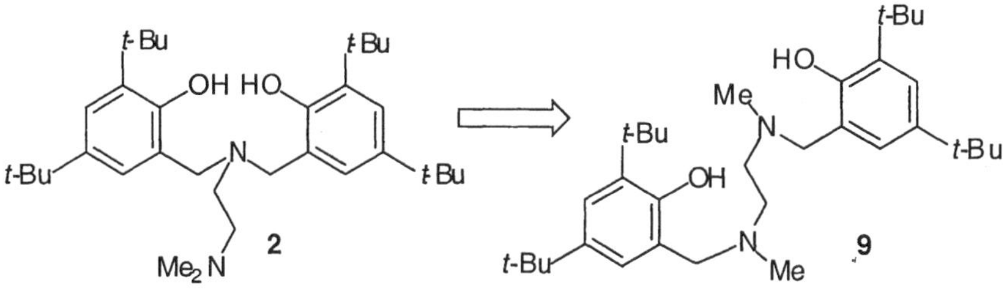

Figure 8. The relationship between amine diphenolate ligands (left) and diamine diphenolate ligands (right).

This new family of dianionic tetradentate chelating ligands may be easily synthesized by the same methodology employed for making the first family of ligands, but employing different starting materials, namely, a one-pot Mannich condensation between the readily available di(secondary) amines, formaldehyde and substituted phenols as demonstrated in Figure 9 (*20*).

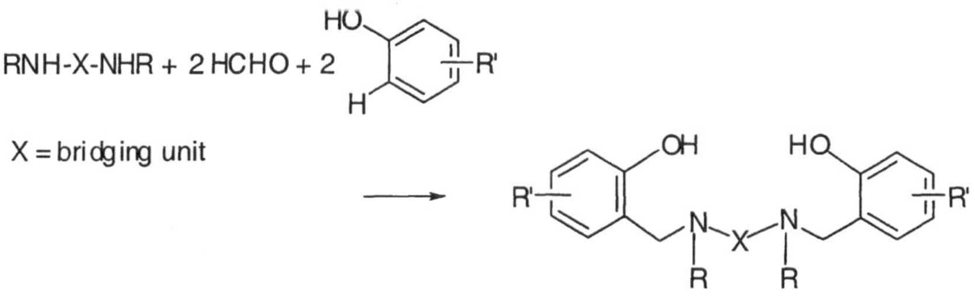

Figure 9. Synthesis of diamine diphenolate ligand precursors.

Zirconium Complexes – Synthesis, Structure and Reactivity

For these ligands too, the most convenient entry into zirconium dialkyl complexes was found to be the reaction between the ligand precursors and tetrabenzyl zirconium at 65 °C. The wrapping of a sequential [ONNO]-type ligand around the metal may lead to several possible isomers as shown in Figure 10.

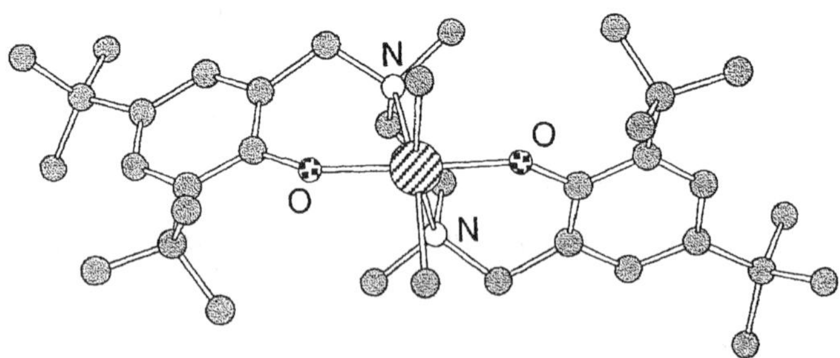

cis(Bn,Bn)-*trans*(O,O) isomers: C_2 or C_1 symmetry

cis(Bn,Bn)-*cis*(O,O) isomers: C_1 symmetry

trans(Bn,Bn) isomers: C_s or C_2 symmetry

Figure 10. Possible octahedral isomers that may form by a sequential [ONNO]-type ligand.

The preferred isomer should have a cis relationship between the benzyl groups for effective catalytic activity, and an overall C_2-symmetry that may induce tacticity independently of the rearrangement rate of the growing polymer chain. The ligand precursor **9** reacted with tetrabenzyl zirconium giving the single isomer **10** in quantitative yield. NMR and X-ray crystallography (Figure 11) indicated that the desired complex was the one formed.

*Figure 11. Molecular structure of **10** exhibiting the C_2 symmetry and cis(Bn,Bn) geometry. H atoms and phenyl groups omitted for clarity.*

Activating the dibenzyl complex **10** with B(C$_6$F$_5$)$_3$ in neat 1-hexene at room temperature led to a mild polymerization catalyst (18 g mmol cat^{-1} h^{-1}, after 30 min). Olefinic termination groups were not observed in the ^{1}H NMR spectra, and the polymer obtained had a molecular weight of M_w = 12,000 and a narrow

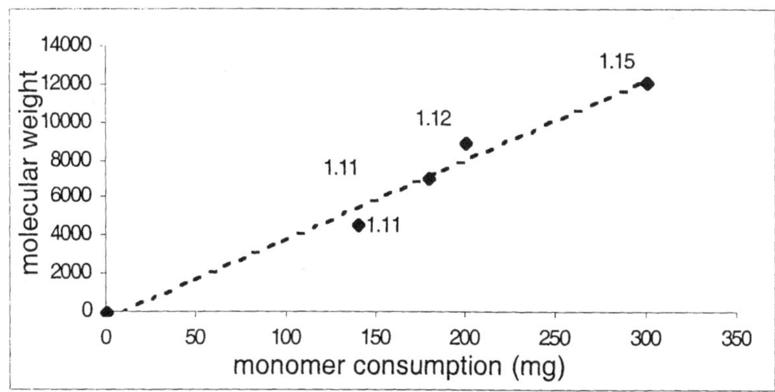

Figure 12. Dependence of M_w on monomer consumption using **10** *activated with* $B(C_6F_5)_3$.

PDI of M_w/M_n = 1.15. The narrow PDI and the linear dependence of the polymer molecular weight on the consumption of the monomer shown in Figure 12 suggest that the polymerization system is living.

Furthermore, six narrow singlets in the ^{13}C NMR indicate an isospecific 1-hexene polymerization, affording >95% isotactic poly(1-hexene). This catalyst is therefore the first to afford the combination of living *and* isotactic polymerization of alpha-olefins at room-temperature.

A possible source of this high isotacticity is the bulk of the *tert*-butyl groups which direct the approaching olefin. To evaluate the effect of steric bulk on the tacticity, we synthesized a ligand precursor featuring methyl substituents on the phenolate rings. The corresponding zirconium dibenzyl complex was obtained quantitatively as a single isomer of the same C_2-symmetry as **10**. Under the same polymerization conditions the less bulky complex was somewhat more active (35 g mmol cat^{-1} h^{-1}), however, the poly(1-hexene) produced from it was atactic. Thus, the size of the substituents was shown to have a significant influence on the tacticity of the poly(1-hexene) produced using these C_2-symmetrical sequential [ONNO]Zr-type catalysts (*21*).

A variety of ligand precursors and metal complexes may be obtained using the methodology outlined above. We are currently developing new catalysts of this family.

Conclusion

Systematic variations of basic parameters of the amine phenolate ligands, including the number and type of donor atoms, steric bulk, and connectivity mode, combined with variations of the metal enable a close control of the structure of the resulting group IV metal complexes. The activity of the resulting complexes was found to be strongly dependent on these structural changes. Thus, a variety of high olefin polymerization catalysts were introduced, including: extremely active catalysts, living polymerization catalysts, and isotactic polymerization catalysts. Further structure-activity studies of these systems and related ones are currently under way.

Acknowledgements

This research was supported in part by the Israel Ministry of Science, Culture and Sports, and in part by the Israel Science Foundation, founded by the Israel Academy of Sciences and Humanities.

References

1. Britovsek, G. J. P.; Gibson, V. C; Wass, D. F. *Angew. Chem. Int. Ed. Engl.* **1999**, *38*, 428.
2. Scollard, J. D.; McConville, D. H. *J. Am. Chem. Soc.* **1996**, *118*, 10008.
3. Baumann, R.; Davis, W. M.; Schrock, R. R. *J. Am. Chem. Soc.* **1997**, *119*, 3830.
4. Matsui, S.; Mitani, M.; Saito, J.; Tohi, Y.; Makio, H.; Matsukawa, N.; Takagi, Y.; Tsuru, K.; Nitabaru, M.; Nakano, T.; Tanaka, H.; Kashiwa, N.; Fujita, T. *J. Am. Chem. Soc.* **2001**, *123*, 6847.
5. Tian, J.; Hustad, P.D.; Coates, G. W. *J. Am. Chem. Soc.* **2001**, *123*, 5134.
6. Mitani, M.; Furuyama ,R.; Mohri, J.; Saito, J.; Ishii, S.; Terao, H.; Kashiwa, N.; Fujita, T. *J. Am. Chem. Soc.* **2002**, *124*, 7888.
7. Hirotsu, M.; Kojima, M.; Yoshikawa, Y. *Bull. Chem. Soc. Jpn.* **1997**, *70*, 649.
8. Zurita, D.; Menage, S.; Pierre, J. L.; Saint-Aman, E. *New J. Chem.* **1997**, *21*, 1001.
9. Tshuva, E. Y.; Versano, M.; Goldberg, I.; Kol, M.; Weitman, H.; Goldschmidt, Z. *Inorg. Chem. Commun.* **1999**, *2*, 371.
10. Burke, W. J.; Glennie, E. L. M.; Weatherbee, C. *J. Org. Chem.* **1964**, *28*, 909.
11. Tshuva, E. Y.; Goldberg, I.; Kol, M.; Weitman, H.; Goldschmidt, Z. *Chem. Commun.* **2000**, 379.
12. Tshuva, E. Y.; Goldberg, I.; Kol, M.; Goldschmidt, Z. *Organometallics* **2001**, *20*, 3017.
13. Tshuva, E. Y.; Groysman, S.; Goldberg, I.; Kol, M.; Goldschmidt, Z. *Organometallics* **2002**, *21*, 662.

14. Tshuva, E. Y.; Goldberg, I.; Kol, M.; Goldschmidt, Z. *Inorg. Chem.* **2001**, *40*, 4263.

15. Groysman, S. *Unpublished Results.*

16. Horton, A. D.; de With, J. *Organometallics* **1997**, *16*, 5424.

17. Tshuva, E. Y.; Goldberg, I.; Kol, M.; Goldschmidt, Z. *Inorg. Chem. Commun.* **2000**, *3*, 610.

18. Tshuva, E. Y.; Goldberg, I.; Kol, M.; Goldschmidt, Z. *Chem. Commun.* **2001**, 2120.

19. Tshuva, E. Y.; Goldberg, I.; Kol, M. *J. Am. Chem. Soc.* **2000**, *122*, 10706.

20. Tshuva, E. Y.; Gendeziuk, N. Kol, M. *Tetrahedron Lett.* **2001**, *42*, 6405.

21. For tactic polymerization of propylene using our catalysts, see: Busico, V.; Cipullo, R.; Ronca, S.; Budzelaar, P. H. M. *Macromol. Rapid Commun.* **2001**, *22*, 1405.

Chapter 6

Modeling and Catalytic Performance of Group 4 Metal Complexes with Anionic Heteroatomic Ligands

Sandor M. Nagy[1], Mark P. Mack[2], and Gregory G. Hlatky[2]

[1]Process Research Center, Equistar Chemicals LP, 8935 North Tabler Road, Morris, IL 60450
[2]Cincinnati Technology Center, Equistar Chemicals LP, 11530 Northlake Drive, Cincinnati, OH 45249

The catalytic performance of certain early transition metal polymerization catalysts with heteroatomic ligands have been studied using a simple molecular modeling approach. This allows us to rationalize experimental observations and provides guidance in developing new ligands for single-site polymerization catalysts.

Olefin polymerization catalysis has undergone an extraordinary transformation in the last 15 years with the development of metallocene polymerization catalysts. The potential of these catalysts to generate polymers with controllable molecular architectures has led to an enormous investment in research and development by both industrial and academic investigators.

In the last twenty years, polyolefin catalysis has been notable for the progression from heterogeneous, ill-defined catalysts to well-characterized metal complexes with high catalytic activity. "First-generation" metallocene catalysts, discovered a few years after Ziegler-Natta systems (*1*), were useful for gaining insights into fundamental polymerization processes, but the low activity and limited stability of the $Cp_2TiCl_2/AlEt_2Cl$ catalysts made them useful only for laboratory studies.

The combination of the discoveries that methylalumoxane was a highly effective cocatalyst (*2*) and that substitution on the cyclopentadienyl rings could influence molecular weight, activity, comonomer incorporation, and stereospecificity (*3*) unleashed the commercial potential of metallocene catalysts. This has led to the preparation of a remarkable variety of metallocenes to control the characteristics of the polyolefin. Some of the polymers produced from these catalysts -- syndiotactic polypropylene and polystyrene and hemiisotactic polypropylene -- are inaccessible using classic Ziegler-Natta systems.

The development of borate-based activators (*4*) to replace methylalumoxane has given us ion-pair catalysts with activities as high as their alumoxane-based counterparts. *These same catalysts* also have discrete formulations which permit a comprehensive characterization of the active polymerizing species, thus bringing catalyst formulation and control to unprecedented levels of understanding while still having commercially acceptable activities.

The C_5H_5 group and its substituted derivatives are ancillary ligands; that is, ligands not directly involved in the catalytic process. Their function is twofold: 1) controlling the steric and electronic nature of the polymerizing site; 2) preventing the formation of secondary active sites which broaden MWD and CD.

A huge number of metallocene procatalysts have been developed over the last 15 years, but a question remains: is the cyclopentadienyl group a *sine qua non* for effective single-site catalysis? So long as the ancillary ligand effectively prevents formation of multiple active sites and remains stable in the course of polymerization, there is no inherent reason why the ligand set must be limited to cyclopentadienyl groups. Thus "metallocene" olefin polymerization catalysts -- catalysts with metals having one or two cyclopentadienyl groups -- are only a subset of a potentailly much wider field of "single-site" catalysts, which may not be metallocene complexes at all. The popularity of cyclopentadienyl-bearing complexes in olefin polymerization catalysts may stem from nothing more than a historical coincidence of two serendipitous discoveries dating from the 1950's: the metal-catalyzed polymerization of olefins and the preparation of cyclopentadienyl metal complexes, starting with ferrocene.

Isolobal groups are one of the subclasses of ancillary ligands which are alternatives to cyclopentadienyl groups in the forging of novel single-site catalysts. A wide range of such alternative ligands, including tris(pyrazolyl)borates (**1**) (*5*), pyrroles and carbazoles (**2**) (*6*), phospholes (**3**) (*7*), boratabenzenes (**4**) (*8*), azaborolines (**5**) (*9*), and thiaborolines (**6**) (*10*), have been used in Group 4 metal complexes serving as single-site catalysts.

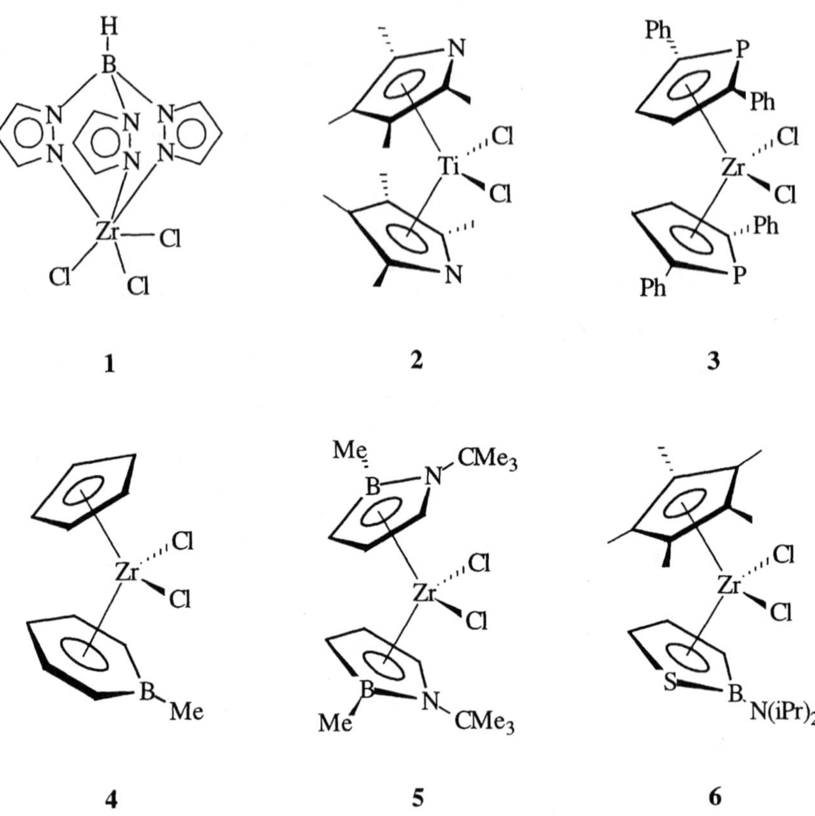

An exciting feature of such heteroaromatic π-ligands is the extremely broad range of electronic properties they exhibit, far greater than the simple alkyl substitutions on cyclopentadienyl rings. This is illustrated in Table 1, which presents some computational values of the acidities of precursors of some of the heteroaromatic anions. A hybrid DFT (B3LYP, TITAN Program) method was used to calculate the enthalpy of the reaction in which the anion is protonated, without taking solvatation into the account. This enthalpy can be used as a measure of relative gas phase acidities of the corresponding precursors.

The precursors of methylboratabenzene anion and its benzo-annealed analogues are more acidic than cyclopentadiene and its substituted derivatives, while the azaborole precursors are significantly less acidic. Obviously, substitution of the methyl group on the boron with the strongly electron-donating amino-group, significantly reduces the acidity of these precursors. Even though the acidity in solution is significantly affected by solvation effects, experimental results confirm these computational estimates. If methylboratacyclohexadiene

Table I. Calculated Gas Phase Basicities of Aromatic and Heteroaromatic Anions Relative to Cyclopentadienyl Anion.

Anion	$\Delta\Delta H$ (kcal/mole)
Benzyl	28
B-Me-N-Me-1,2-azaborolinyl (Me$_2$-AB)	14.3
B-NMe$_2$-1,2-thiaborolinyl (Me$_2$N-TB)	8.2
Cyclopentadienyl (Cp)	0
B-Me-1,2-thiaborolinyl (Me-TB)	-3.5
Indenyl (Ind)	-5.1
Fluorenyl (Flu)	-6.6
B-NMe$_2$-2-boratanaphthalene (Me$_2$N-BN)	-9.1
B-NMe$_2$-boratabenzene (Me$_2$N-BB)	-10.8
B-Me-9-borataanthracene (Me-BA)	-20.3
B-Me-2-boratanaphthalene (Me-BN)	-21.4
B-Me-boratabenzene (Me-BB)	-22.3
B-Me-borataphenanthrene (Me-BN)	-22.3

can be easily deprotonated by sodium hydride, the azaborole precursor require much stronger deprotnating agents, like n-BuLi/KO-t-Bu. These calculated acidities correlate extremely well with the gas-phase experimental values determined by Ashe (*11*) for methylboratacyclohexadiene.

Considering that the cyclopentadienyl and its heteroaromatic analogues are donor ligands, it is quite informative to rank them according to some computational indices that reflect their donor properties. The energy of the highest molecular orbital and the chemical softness, characterized as the HOMO-LUMO gap are such convenient parameters. Better donors have higher HOMOs and are more soft or more polarizable and as a result, more efficiently responding to a demand from an electron acceptor center such as the metal center.

Figure 1 is a plot of the HOMO energy and chemical softness of cyclopentadienyl derivatives and several of heteroaromatic analogues. The expected donating power of ligands in general increases on going from the methylboratabenzene series to the amino-boratabenzene and Cp series to that of azaboroles. The softness in each of these series increases as an annelated benzo ring is attached. What is also clear is that these heteroaromatic ligands provide a significantly *broader* spectrum of donor properties compared to the traditional Cp-type ligands expanding the options available for catalyst design.

The relative stability of the highly electrophilic active sites determines how easy it will be to activate the precursor with the cocatalyst or what proportion of the metal centers are present in active form. This, besides the inherent activity of

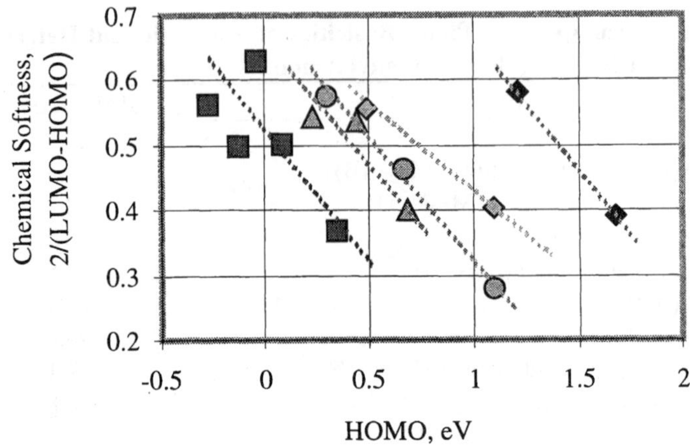

Figure 1: HOMO energies and chemical softness of anionic ligands (● = Cp series; ■ = Me-boratabenzene derivatives; ▲ = dimethylaminoboratabenzene derivatives; ♦ = azaborolinyl derivatives)

the sites, to a significant extent determines the activity of the catalytic system. The stability of the cationic active sites is directly related to the donor ability of the ligands. We found that the calculated stabilization correlates well with the ligand's softness.

Figure 2 plots the calculated stabilization of the methyl zirconocenium ion relative to the bis-cyclopentadienyl cation. We can see that the softer ligands with extended π-systems have the strongest stabilizing effect and most of the heteroaromatic ligands fall on the same trend line as the cyclopentadienyl series. Remarkably, the amino-boratabenzene series have a slightly enhanced stabilizing effect compared to other ligands examined.

The identity of the ligand influences not only the relative stability of active sites, but also their inherent reactivity toward ethylene, comonomer or chain transfer agents. We found that this reactivity correlates reasonably well with the LUMO energy of the cation, what makes a perfect sense for an orbitally controlled process. This means that we can use the LUMO energy as a simple reactivity index to estimate the relative reactivity of active sites with different ligands: the lower the orbital energy the higher the expected reactivity of the site toward ethylene or other Lewis bases.

Figure 3 shows the LUMO energies of zirconocenium ions with different ligands against the softness of the active sites for series of cyclopentadienyl, methyl- and aminoboratabenzene and azaborole ligands. Again there is a trend of

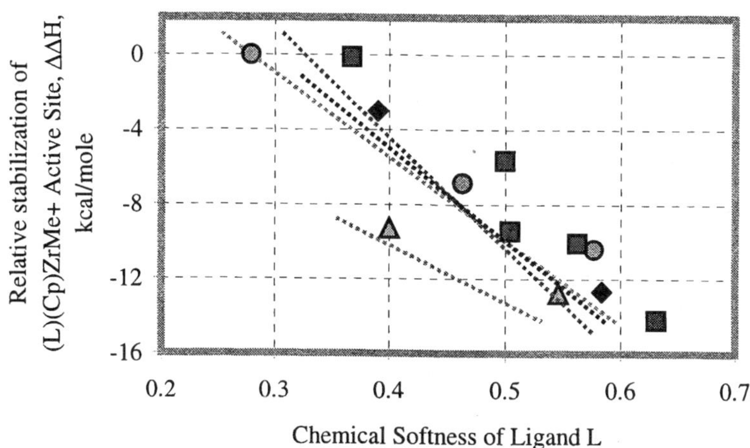

Figure 2: Correlation of ligand softness with relative stabilization in (L)CpZrMe⁺ species (● = Cp-type anions; ■ = methylboratbenzene derivatives; ▲ = dimethylaminoboratabenzene derivatives; ♦ = 1,2-azaborole derivatives).

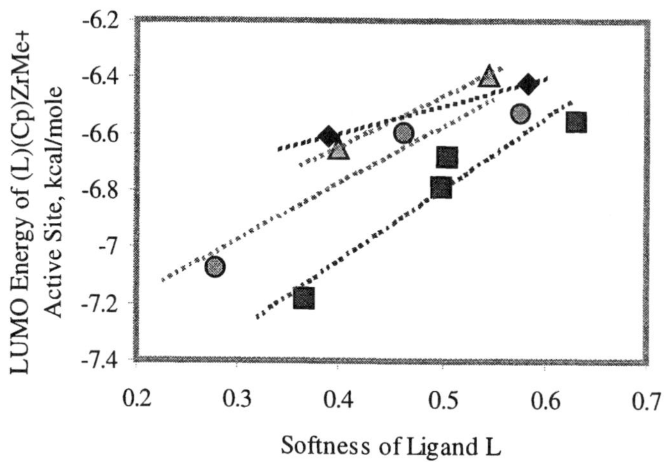

Figure 3: Correlation of ligand softness with LUMO energy in (L)CpZrMe⁺ species (legends as in Figure 2).

increased expected inherent activity, associated with the lower LUMO energy, as the softness of the ligand diminishes. In this case the methyl-boratabenzene series exhibit enhanced reactivity at same softness level.

We studied the polymerization performance of a number of complexes to validate these concepts. These zirconium complexes had one ligand varied from cyclopentadienyl to methylboratabenzene to (diisopropylamino)boratabenzene and the corresponding 2-boratanaphthalenes. The catalysts were supported on silica and tested under conditions close to those used in commercial slurry-loop polymerization processes. The kinetic profile of all processes, as measured by ethylene consumption was similar, characterized by a high initial rate that slightly declined during the process.

In our experiments the methylboratabenzene-based complex had the highest activity. The activity increases as the LUMO energy calculated for the methyl zirconocenium model decreases (Figure 4), indicating that there is a similar level of active sites and that the LUMO energy is a reasonably good reactivity index for rough prediction of catalyst activity.

When ethylene was copolymerized with large levels of butene comonomer (*12*) there was a large difference in the molecular weights of copolymers

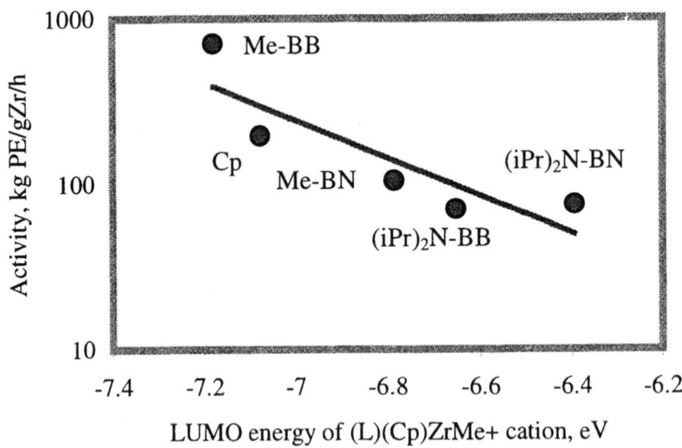

Figure 4: Correlation of LUMO energies and polymerization activities in (L)CpZrMe+ catalysts (Cp = cyclopentadienyl; BB = boratabenzene; BN = 2-boratanaphthalene).

produced with different complexes. If the molecular weight of the polymer is proportional to the ratio of propagation and termination rate constants, then it can be expected that a reactivity index, which accounts for the relative stability of key intermediates leading to propagation and termination, would be a reasonable computational parameter to be used in correlations (eqn. 1). Indeed, this is what is observed (Figure 5). We conclude that the low molecular weights of the co-polymers produced with methylboratabenzene complexes are related to the increased relative stability of β-agostic structures (Figure 6).

$$Mw \sim k_{prop}/k_{term} \sim [Exp(-E_a^{prop})]/[Exp(-E_a^{term})] \sim [Exp(-\Delta\Delta H_\pi)]/[Exp(-\Delta\Delta H_\beta)] \quad (1)$$

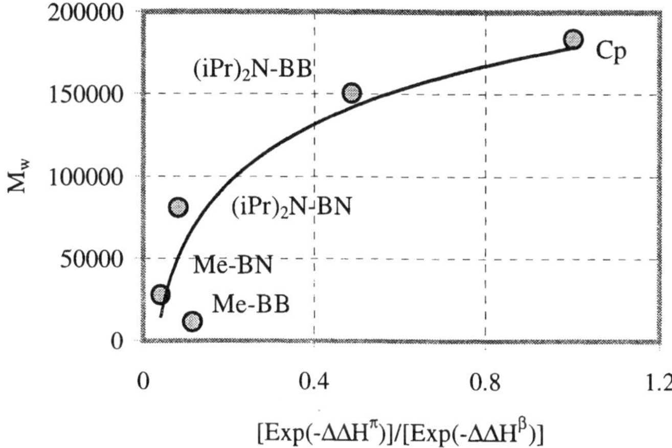

Figure 5: Correlation of calculated ratios of relative heats of π-complexation vs. relative energy of β-agostic stabilization and molecular weight of ethylene/butene copolymers prepared with (L)CpZrMe$^+$ catalysts.

The difference in efficiency of comonomer incorporation observed for these complexes can be rationalized simply by the activity-selectivity principle. A site with higher inherent activity will discriminate to a lesser extent between ethylene and the comonomer resulting in a higher incorporation value. Figure 7 shows the relationship of LUMO energies and ethyl branching in ethylene-1-butene copolymerizations with (L)(Cp)ZrMe$^+$ catalysts. The ethyl branching measured for the polymers prepared with the more active methyl-boratabenzene and cyclopentadienyl complexes is higher than branching observed for the less active complexes.

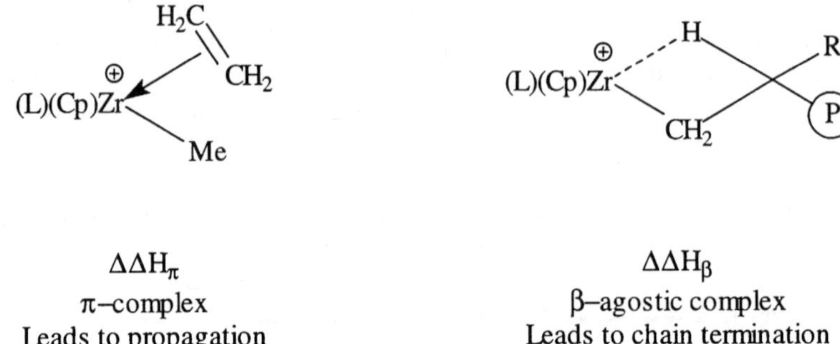

Figure 6: Key intermediates in olefin polymerization.

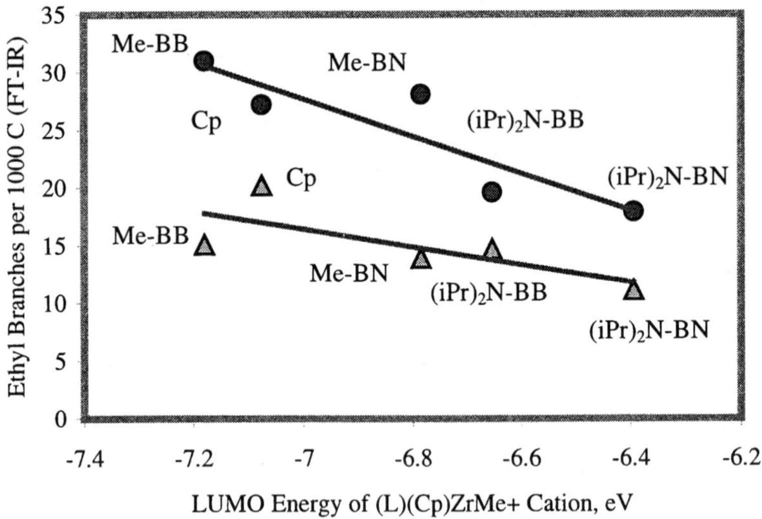

Figure 7: Correlation of LUMO energies and 1-butene incorporation in (L)(Cp)ZrMe⁺ catalysts. (▲ = polymerizations with 100 mL 1-butene, ● = polymerizations with 200 mL 1-butene)

In summary, heteroaromatic anionic ligands are not just viable alternatives to the more traditional cyclopentadienyl-type ligand in transition metal complexes used in olefin polymerization, in addition they broaden considerably the spectrum of electronic and steric effects beyond those seen in substituted cyclopentadienyl-based catalysts.

Ligand modification is a powerful tool for controlling catalytic performance. Molecular modeling allows us to quantify and predict the ligand effect and design the catalyst rationally.

Acknowledgments

We would like to acknowledge the key contribution to this study made by our colleagues Drs. Bradley Etherton and Ramesh Krishnamurti and the management of our companies, Occidental Chemical and Equistar Chemical for support and permission to present these data.

References

1. (a) Natta, G.; Pino, P.; Mazzanti, G.; Giannini, V. *J. Am. Chem. Soc.* **1957**, *79*, 2925. (b) Breslow, D. S.; Newburg, N. R. *J. Am. Chem. Soc.* **1957**, *79*, 5072.
2. Sinn, H.; Kaminsky, W. *Adv. Organometal. Chem.* **1980**, *18*, 99.
3. (a) Welborn, H. C.; Ewen, J. A. U.S. Patent 5,324,800, 1994. (b) Ewen, J. A. *Stud. Surf. Sci. Catal.* **1986**, *25*, 271.
4. (a) Turner, H. W.; Hlatky, G. G., Eckman, R. R. U.S. Patent 5,198,401, 1993. (b) Ewen, J. A.; Elder, M. J. U.S. Patent 5,387,586, 1995. (c) Ewen, J. A.; Elder, M. J. U.S. Patent 5,561,092, 1996.
5. Kohara, T.; Ueki, S. U.S. Patent 4,870,042, 1989.
6. Etherton, B. P.; Nagy, S. U.S. Patent 5,539,124, 1996.
7. de Boer, E. J. M.; Gilmore, I. J.; Korndorffer, F. M.; Horton, A. D.; van der Linden, A; Royan, B. W.; Ruisch, B. J.; Schoon, L.; Shaw, R. W. *J. Mol. Catal. A* **1998**, *128*, 155.
8. (a) Bazan, G. C.; Rodriguez, G.; Ashe, III, A. J.; Al-Ahmad, S.; Müller, C. *J. Am. Chem. Soc.* **1996**, *118*, 2291. (b) Krishnamurti, R.; Nagy, S.; Etherton, B. P. U.S. Patent 5,554,775, 1996
9. Nagy, S.; Krishnamurti, R.; Etherton, B. P. U.S. Patent 5,902,886, 1999.
10. Ashe, III, A. J.; Fang, X.; Kampf, J. W. Organometallics **2000**, *19*, 4935
11. Sullivan, S. A.; Sandford, H.; Beauchamp, J. L.; Ashe, III, A. J. *J. Am. Chem. Soc.* **1978**, *100*, 3737.
12. The zirconium complexes (75 mmoles) were supported on $AlMe_3$-treated silica with $[Ph_3C][B(C_6F_5)_4]$ cocatalyst (75 mmoles). Polymerizations were run in a 2 L autoclave with ethylene (350 psi) and 1-butene comonomer (100 or 200 mL) at 70°C for 30 minutes. Branching analysis was done by FT-IR and molecular weights determined by GPC.

Catalysts from Groups 5 and 6 Metal Complexes

1,3,5-Triazacyclohexane Complexes of Chromium as Homogeneous Model Systems for the Phillips Catalyst

Randolf D. Köhn[1], D. Smith[1], D. Lilge[2], S. Mihan[2], F. Molnar[3], and M. Prinz[3]

[1]Department of Chemistry, University of Bath, Bath BA2 7AY, United Kingdom
[2]Basell Polyolefins GmbH, Carl-Bosch-Strasse M505, D–67056 Ludwisshofer, Germany
[3]BASF Aktiengesellschaft GKT/P B1, 87056 Ludwigshafen, Germany

Complexes of N-substituted 1,3,5-triazacyclohexanes with $CrCl_3$ can be activated by MAO or $PhNMe_2H(B(C_6F_5)_4)/AlR_3$ to give solutions that can polymerize and/or trimerize ethylene depending on the N-substituents R with activities and polymer products similar to those of the heterogeneous Phillips catalysts. α-olefins are selectively trimerized or co-polymerized with ethylene. Variation of these substituents R showed a large dependence of the trimerization/polymerization ratio on branching in the N-substituent. Spectroscopic studies show that the triazacyclohexane stays co-ordinated during the catalysis and that mono-nuclear metallacyclic complexes with a weakly coordinating anion in one coordination site are likely involved.

Introduction

Several transition metal systems have been found over the past 50 years that can catalyze the polymerization and oligomerization of olefins. Heterogeneous Ziegler-Natta systems based on early transition metals are the most successful and produce a large variety of polyolefins. Homogeneous single site model systems, foremost the metallocenes, have been developed which are able to produce highly stereo regular polymers and have become useful industrial catalysts. The study of these homogeneous systems has tremendously improved the detailed understanding of the mechanism. Chain propagation is largely based on olefin insertion into metal-C or H bonds followed by chain transfer via β-H elimination. This general mechanism is often termed the hydride mechanism and seems to be common among other transition metal systems as well.

Late transition metal systems often have a higher tendency for chain transfer versus propagation that leads to good dimerization or oligomerization catalysts. One highly successful system is the nickel based SHOP catalyst for the ethylene oligomerization. Blocking the chain transfer pathway with sterically demanding groups in similar systems has lead to an exciting development of highly active homogeneous polymerization catalysts based on late transition metals such as Fe, Co, Ni, and Pd. Again, a hydride mechanism appears to be involved.

Chromium based Catalysts

Polymerization

The heterogeneous Phillips catalysts (*1,2,3,4,5*) based on CrO_3/SiO_2 for the polymerization of ethylene have been known as long as the Ziegler-Natta systems and still produce a large fraction of the world production of HDPE (>7 million t/a)(*6,7*). This system has many unusual features compared to the other transition metal catalysts. First of all, it is a purely inorganic system that does not require any metal alkyl co-catalyst and can be activated by ethylene itself although activation can also be achieved with other reducing reagents such as CO or aluminum alkyls. Contrary to most catalysts based on the hydride mechanism, the molecular weight of the polymer is quite insensitive to hydrogen but can be regulated by the reaction temperature. End group analysis of the polymer shows not only the expected single methyl and vinyl end groups but additional methyl groups, vinylidene and some internal olefin end groups. Such end groups have been found in systems based on the hydride mechanism when additional isomerisation, mis-insertion or chain-walking steps are involved.

However, these groups are found in Phillips systems in consistently similar ratios. The Phillips catalysts show characteristically broad molecular weight distributions. Previous studies have shown that several different chromium sites are formed during the calcinations and reduction steps and that only a small fraction of these are actually active.

When activated with metal alkyls the Phillips catalyst can also oligomerize ethylene to α-olefins that can be in-situ co-polymerized giving polymer with side chains. However, these oligomers do not follow the Flory-Schultz distribution typical for systems based on the hydride mechanism. There is a high selectivity for the trimer of ethylene, 1-hexene, and subsequent co-polymers with butyl side chains. Co-polymers can also be obtained directly by adding α-olefins.

Contrary to the early and late transition metal systems, the nature of the active species and the mechanism in the chromium systems is still a matter of debate. This is largely due to the fact that the Phillips catalyst is very difficult to study. Highly active catalysts are obtained only at high dilution of chromium on the silica surface (< 1 w%) and the chemistry of the surface chromium is very rich with many different species at various oxidation states and nuclearities. However, only a small fraction of the total chromium is known to be active. Thus, surface analytical methods mostly detect inactive compounds. Generally, a mononuclear chromium compound in the oxidation state +III directly bound to the silica via two or three oxygen atoms is believed to be the active species (Figure 1). How additional Cr-C bonds are formed during catalysis is still unclear.

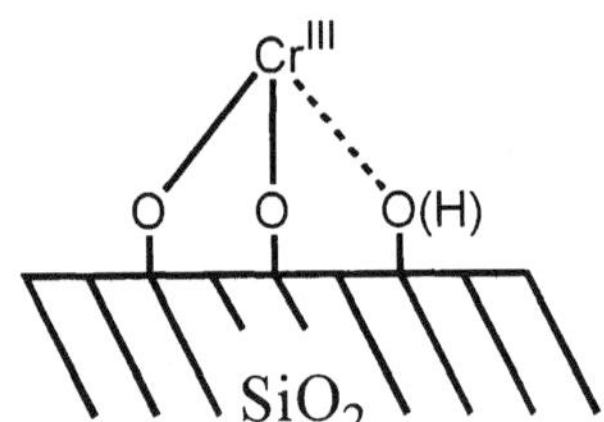

Figure 1. Possible co-ordination environment of the active site in the Phillips catalyst

Many homogeneous model systems (*8*) show only limited activity (*9,10*) and most of them fail to reproduce the properties of the Phillips catalyst and a true model has yet to be found. As an alternative chromium system, the Union Carbide catalysts based on chromocene on silica has been introduced 30 years ago which is more accessible to studies. One cyclopentadienyl appears to stay attached to chromium and various mono-cyclopentadienyl chromium complexes show polymerization activity. However, these cyclopentadienyl systems are quite different from the Phillips catalyst. They generally do not co-polymerize ethylene and α-olefins, the molecular weight of the polymer shows high hydrogen response, the polymers do not have the same end group distribution and are mostly linear and there is no selectivity for trimerization.

Ethylene Trimerization

In an interesting development after the initial report by Briggs in 1987 (*11*), homogeneous chromium systems have been found that can trimerize ethylene with >90% selectivity to 1-hexene. These systems generally consist of some soluble chromium complex, aluminum alkyls and some amine, mostly pyrroles (*12*) or more recently PNP ligands (*13*). The increasing demand in 1-hexene has sparked a growing interest in these selective trimerization catalysts as can be seen in the number of publications in the past few years (Figure 2).

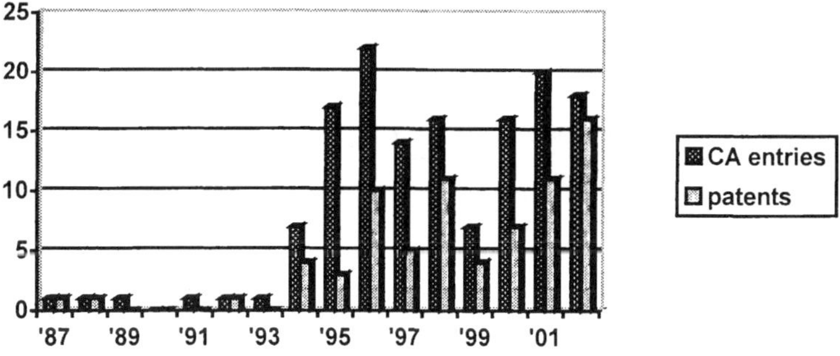

Figure 2. Chemical abstract entries on trimerization of ethylene

However, little is known about the active species in these complex mixtures largely due to the difficulty of obtaining useful NMR spectra of these highly paramagnetic compounds. Briggs (*11*) and others have proposed a mechanism via metallacycles analogous to Figure 8 and Jolly (*14*) was able to support this by showing that complex shown in Figure 3 can react with ethylene under reducing conditions (activated Mg) to a metallacyclopentane complex and that the larger metallacycloheptane complex readily decomposes under reductive elimination to the trimer 1-hexene. However, insertions into the Cr-C bonds are prevented by the amine donor and addition of MAO leads to an active ethylene polymerization catalyst without trimerization probably by opening the the metallacycle.

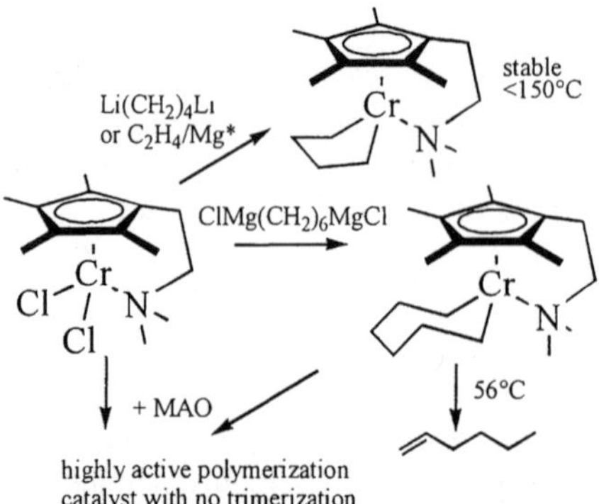

Figure 3. Formation and decomposition of metallacycles.

Apparently, a reduced chromium species and oxidative addition of two olefins to a metallacyclopentane is involved. Only after insertion of a third olefin, resulting in a metallacycloheptane, reductive elimination becomes possible. Thus, oligomerization is selectively terminated after three olefins. This means that a new mechanism for the C-C bond formation between olefins and metallacycles are involved in the selective trimerization. A similar mechanism has recently been found for titanium systems (*15,16*). Thus, a true model system for the Phillips catalyst should be a well defined chromium complex that is also capable of trimerizing ethylene.

Catalysts based on 1,3,5-Triazacyclohexane Chromium Complexes

For several years we have been investigating the co-ordination chemistry of N-substituted 1,3,5-triazacyclohexanes (R₃TAC) (*17-21*). The three hard donor nitrogen atoms can facially co-ordinate metals and may model the two to three oxygen atoms believed to be bound to chromium in the Phillips catalyst. The few investigations on larger triazacycloalkanes, mainly triazacyclononanes, indicate that the N-substituted ligands are sterically too demanding to allow high activity for the polymerization of ethylene or even α-olefins. Our results on the co-ordination chemistry of R₃TAC show that complexes with low steric demand by the ligand due to the small N-metal-N angle of 60-65° can be formed. This small angle and the mis-directed nitrogen lone pairs lead to relatively weak M-N bonds. This can be observed by UV/vis spectroscopy and X-ray crystallography of the complexes [R₃TACCrCl₃] (*22*). The observed 10Dq value of 14,000 cm⁻¹ is much lower than expected for [(amine)₃CrCl₃] (17,000 cm⁻¹) and the Cr-N distance of 210-214 pm is longer than expected for Cr(III)-N(amine) (200 pm).

However, these values indicate a comparable if not stronger bond than in the sterically weakened bonds in N-substituted triazacyclononane complexes (*23*). The structural parameters of the co-ordination environment in [R₃TACCrCl₃] do not vary with the bulk of the N-substituent R whereas steric repulsion increases the Cr-N and Cr-Cl distances in the triazacyclononane complex.

We have been able to characterize a few organometallic chromium(III) complexes with R₃TAC and, in the case of zinc, even cationic alkyl complexes have been synthesized. Thus, it should be possible to generate cationic alkyl complexes of chromium which may be able to catalyze the ethylene polymerization. Indeed, [Me₃TACCrCl₃] reacts with methylaluminoxane (MAO) to give a highly active catalyst. However, good solubility of the complexes is crucial for achieving the high productivities. This solubility problem was solved by introducing longer alkyl chains as N-substituents. The solubility in toluene increases dramatically with octyl and dodecyl substituents (*24*). Complexes [R₃TACCrCl₃] can be prepared according to Figure 4 by reacting R₃TAC with [CrCl₃(THF)₃] or more conveniently with CrCl₃ (stored under air) and zinc powder by boiling in toluene under a stream of nitrogen.

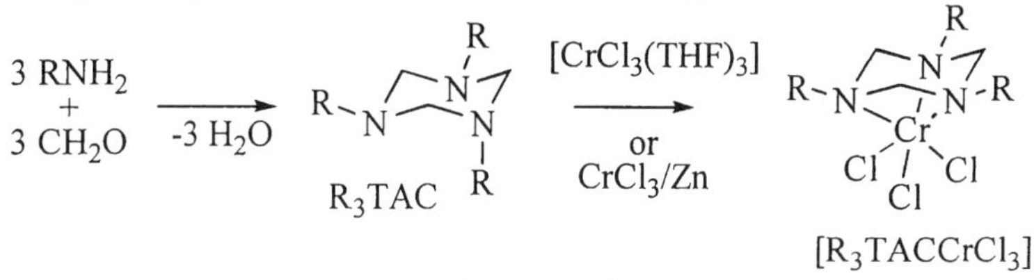

Figure 4. Syntheses of the complexes

Since R₃TAC can also be prepared by heating the corresponding primary amine and paraformaldehyd in toluene, a simple one-pot synthesis is possible (*24*). The complexes are generally inert towards water and air and can be cleaned by chromatography on silica. By using two different amines a mixture of asymmetrically substituted triazacyclohexanes R₂R′TAC can be obtained. If they are not easily separable by distillation or crystallization they can be converted to a mixture of the chromium complexes and separated by chromatography afterwards. This allows not only the synthesis of symmetrically substituted triazacyclohexane complexes but also a large variety of asymmetrically substituted ones including some with functional groups such as ethers, nitriles, amines.

Polymerization

The triazacyclohexane complexes of CrCl₃ can be activated with MAO or [PhNMe₂H⁺] [B(C₆F₅)₄⁻] (DMAB) followed by trialkylaluminum. Activities of up to 800 Kg molCr^{-1}h^{-1} with MAO or up to 2800 Kg molCr^{-1}h^{-1} with DMAB/Al^{i}Bu$_3$ can be obtained. This is better than the best co-catalyst free active complex (*25*) and the best non-Cp-system (*26*). Thus, the activity is comparable to that of the heterogeneous Phillips catalyst (500-3000 Kg molCr^{-1}h^{-1}) and more active chromium catalysts have only been obtained with systems similar to those in Figure 3 (*10,27*) (up to 20,000 Kg molCr^{-1}h^{-1}).

Maximum activity for MAO as co-catalyst is reached at Al:Cr ≈ 300 as seen in Figure 5. However, activation of [R$_3$TACCrCl$_3$] can also be achieved with at least 1 eq. DMAB and 20-50 eq. Al^{i}Bu$_3$ giving similar or better activity.

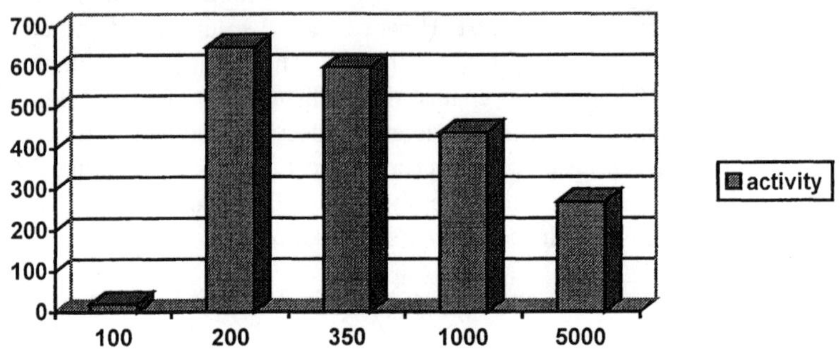

Figure 5. Dependence of the activity on the Al:Cr ratio (for (dodecyl$_3$TAC)CrCl$_3$/MAO/toluene at 40°C/1 bar ethylene)

The activity depends strongly upon the N-substituents R. Figure 6 gives the activities for selected complexes with different N-substituents. Generally, the activity improves with the higher solubility of the long chain substituents and decreases with increasing steric bulk. The molecular weights of the polymers under these conditions are around 40,000 (M$_w$) with M$_w$/M$_n$=2-4 which is typical for a single-site catalyst. Similar to the Phillips catalyst, molecular weights of polyethylene produced by activated [R$_3$TACCrCl$_3$] are highly dependent on the polymerization temperature.

End group analysis of the polymers (IR, ^{13}C-NMR) shows more methyl groups than one expected terminal CH$_3$ and additional vinylidene and some internal olefin besides the expected vinyl end groups in a distribution that is typical for end groups produced by the Phillips catalyst. Since the molecular weights of the polymers differ, the end groups are best compared relative to the total number of olefinic groups (set to 100%) (Figure 7).

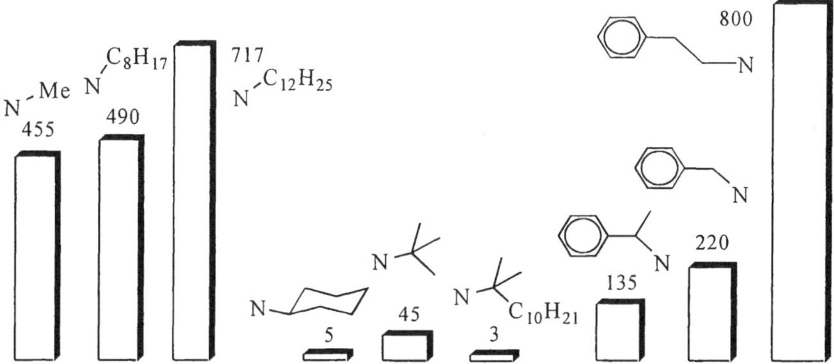

Figure 6. Polymerization with MAO at 40°C in toluene with Al :Cr (300:1) (activity in kg PE/(molCr h))

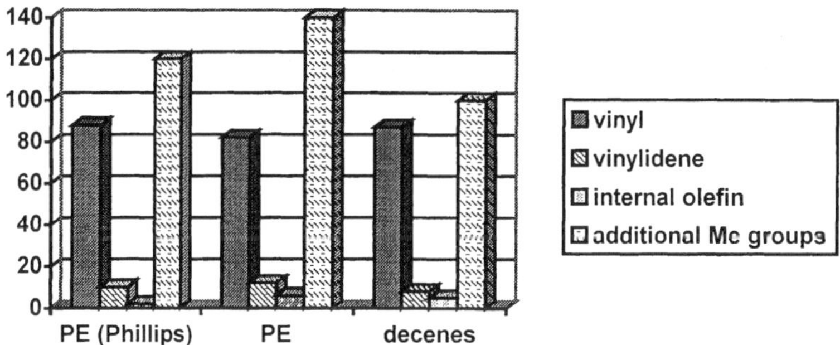

Figure 7. End group distribution relative to the sum of olefinic groups set to 100% (Phillips: typical numbers from industrial polymers produced under various conditions)

Trimerization

Besides polyethylene 1-hexene as the trimer of ethylene and some decenes as "co-trimers" of 1-hexene and ethylene can be found in the solution and butyl side chains in the polyethylene are indicative of some 1-hexene built into the polymer. α-Olefins react with the catalyst system nearly exclusively to the corresponding trimers (22). The homogeneous catalysis of this α-olefin trimerization can be observed by UV/vis and NMR spectroscopy. A single mononuclear chromium(III) complex is observed during the catalysis, probably a metallacyclopentane complex. Two broadened and shifted signals in ^{2}H NMR studies with a ring-deuterated ligand proves that the triazacyclohexane is co-ordinated to the paramagnetic chromium in the active complex. The kinetic data can be fitted to rate laws derived from the metallacyclic mechanism outlined in Figure 8.

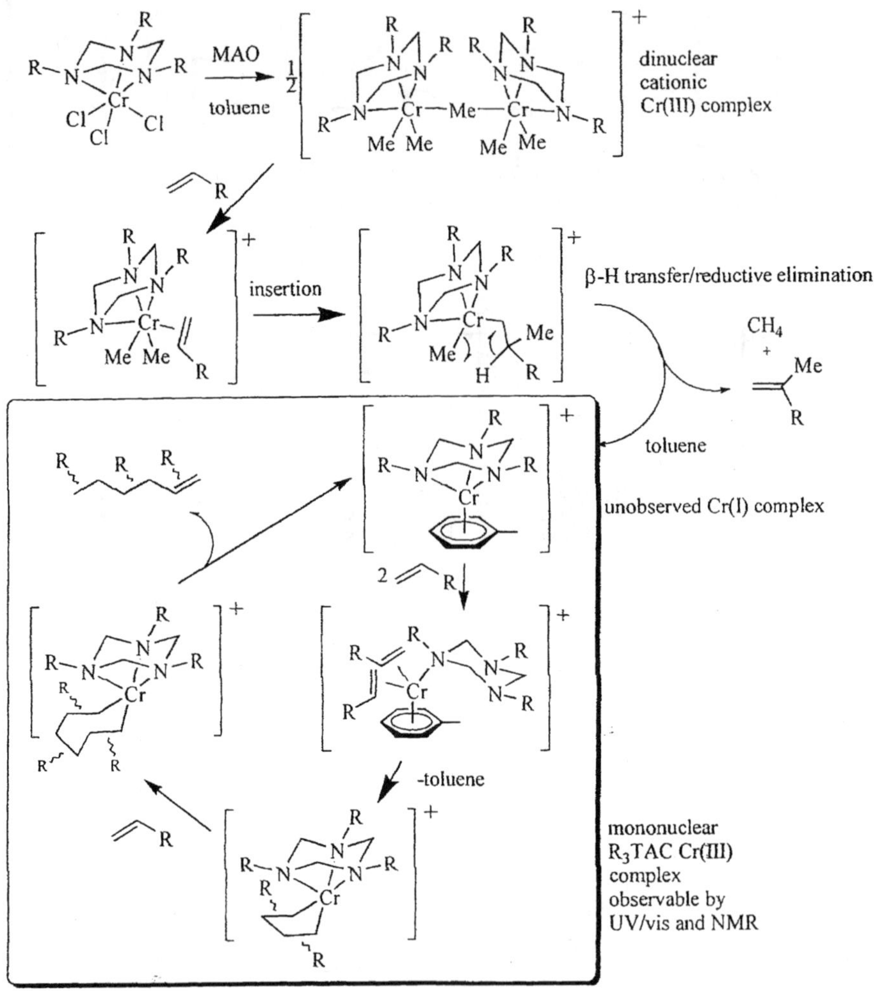

Figure 8. Proposed mechanism for the selective trimerization

Thus, our system is also able to reproduce the selectivity for trimerization. Interestingly, analysis of the decene isomers by NMR shows an end group distribution similar to that in the polymers (Figure 7). Addition of 1-hexene to the solution increases the content of butyl side chains substantially and a co-polymer can be obtained. All observed isomers of the "co-trimers" can be explained by the metallacyclic mechanism according to Figure 8. A similar scheme for the homo-trimerization of α-olefins also yields all observed isomers and supports the formation of these products by the metallacyclic mechanism.

The observed end groups of the polymer could be due to several steps of insertions, eliminations and re-insertions in 1,2 and 2,1 fashion. However, the qualitative and quantitative similarity of the end group distribution between the polymer and the "co-trimer" suggests that the same mechanism via metallacycles is involved with R′ being the polymer chain – at least for the end group formation.

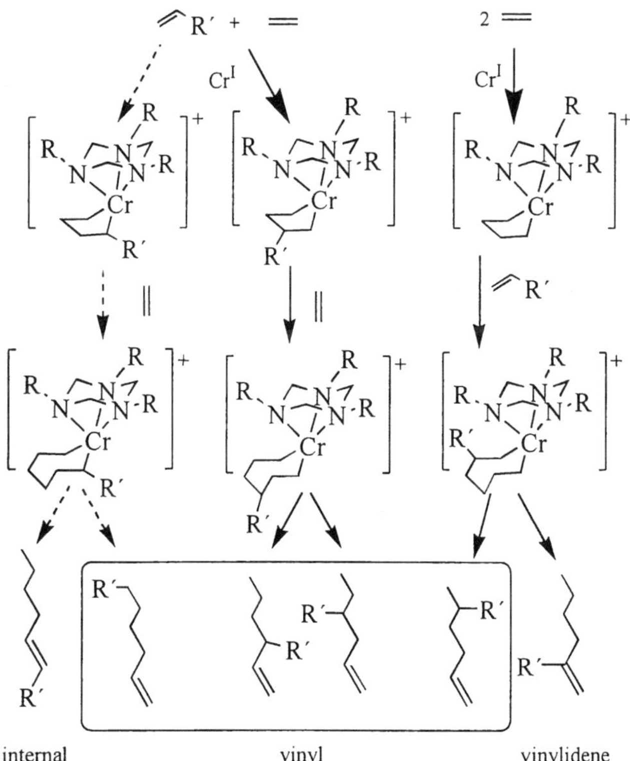

Figure 9. *Origin of the end groups in the product of "co-trimerization" of R′CH=CH₂ and two ethylene.*

The ratio of polyethylene:1-hexene depends mainly on the N-substituent and the reaction temperature. While the complex with N-substituent R=dodecyl gives nearly exclusively polyethylene at ambient conditions, R= 2-ethylhexyl (or other 2-branched substituents) give nearly quantitatively 1-hexene and co-trimers at 50°C. DFT calculations show that the adduct formation of ethylene to the metallacyclopentane (leading to insertion and polymerization) is much easier in the case of R= octyl (8 kJ/mol) than for R= 2-propylheptyl (20 kJ/mol). Thus, 2-branched substituents have a much slower ethylene insertion beyond the trimer stage which allows reductive elimination.

The activation by DMAB/Al^{i}Bu$_3$ can be followed by NMR spectroscopy. Addition of DMAB to [R$_3$TACCrCl$_3$] leads to the disappearance of NMR signals for the aniline and broadening of the ^{19}F NMR signals of the anion. This indicates a close contact to the paramagnetic chromium as a result of an anilinium adduct and ion pairing. Addition of the aluminum alkyl leads to the reappearance of the aniline signals and disappearance of the m-F and severe broadening of the o- and p-F signals. This is also maintained during catalysis. Thus the anion fills a free coordination site in the resting state of the activated catalyst (Figure 10).

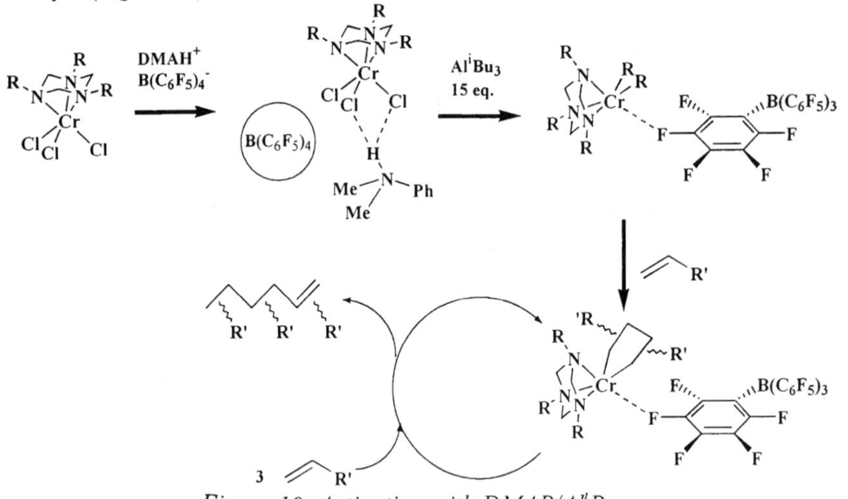

Figure 10. Activation with DMAB/Al^{i}Bu$_3$

Thus, activated [R$_3$TACCrCl$_3$] represents the first true homogeneous model for the Phillips catalyst that can reproduce many important properties and the results indicate that the typical end group distribution may be closely linked to a required trimerization activity.

R. D. Köhn and D. Smith thank EPSRC for support. The authors thank BASF AG and Basell for encouragement and support.

References

1. Hogan, J. P.; Banks, R. L. (Phillips Petroleum) U.S. Patent 2,825,721, 1958
2. McDaniel, M. P. *Adv. Catal.* **1985**, *33*, 47
3. Marsden, C. E. *Plast. Rubber Compos. Process. Appl.* **1994**, *21*, 193
4. Nowlin, T. E. *Prog. Polym. Sci.* **1985**, *11*, 29
5. Augustine, S. M.; Blitz, J. P. *J. of Catalysis* **1996**, *161*, 641.
6. Rätzsch, M. *Kunststoffe* **1996**, *86*, 6
7. Messere, R.; Noels, A.F.; Dournel, P.; Zandona, N.; Breulet, J. *Proceedings of Metallocenes'96*, 6-7 März **1996**, Düsseldorf, 309
8. Theopold, K. H. *Eur. J. Inorg. Chem.* **1998**, *1998*, 15
9. Britovsek, G. J. P.; Gibson, V. C.; Wass, D. F. *Angew. Chem.* **1999**, *111*, 448; *Angew. Chem., Int. Ed. Engl.* **1999**, *38*, 428
10. Döhring, A.; Göhre, J.; Jolly, P. W.; Kryger, B.; Rust, J.; Verhovnik, G. P. J. *Organometallics* **2000**, *19*, 388
11. Briggs, J. R. *J. Chem. Soc., Chem. Commun.*, **1989**, 674
12. Lashier, M. E. (Phillips Petroleum Company) EP Patent 0780353, 1995
13. Carter, A.; Cohen, S. A.; Cooley, N. A.; Murphy, A.; Scutt, J.; Wass, D. F. *J. Chem. Soc., Chem. Commun.* **2002**, 858
14. Emrich, R.; Heinemann, O.; Jolly, P. W.; Krüger, C.; Verhovnik, G. P. J. *Organometallics* **1997**, *16*, 1511
15. Pellecchia, C.; Pappalardo, D.; Oliva, L.; Mazzeo, M.; Gruter, G.-J. *Macromolcules* **2000**, *33*, 2807
16. Deckers, P. J. W.; Hessen, B.; Teuben, J. H. *Angew. Chem.* **2001**, *113*, 2584; *Angew. Chem., Int. Ed. Engl.* **2001**, *40*, 2516
17. Haufe, M.; Köhn, R. D.; Kociok-Köhn, G.; Filippou, A. C. *Inorg. Chem. Commun.* **1998**, *1*, 263
18. Köhn, R. D.; Kociok-Köhn, G. *Angew. Chem.* **1994**, *106*, 1958; *Angew. Chem., Int. Ed. Engl.* **1994**, *33*, 1877
19. Köhn, R. D.; Kociok-Köhn, G.; Haufe, M. *J. Organomet. Chem.* **1995**, *501*, 303
20. Haufe, M.; Köhn, R. D.; Weimann, R.; Seifert, G.; Zeigan, D. *J. Organometal. Chem.* **1996**, *520*, 121
21. Köhn, R. D.; Pan, Z.; Kociok-Köhn, G.; Mahon, M. F. *J.Chem.Soc., Dalton Trans.* **2002**, 2344

22. Köhn, R. D.; Haufe, M.; Kociok-Köhn, G.; Grimm, S.; Wasserscheid, P.; Keim, W. *Angew. Chem.* **2000**, *112*, 4519; *Angew. Chem. Int. Ed. Engl.* **2000**, *39*, 4337
23. Wu, S.-J.; Stahly, G. P.; Fronczek, F. R.; Watkins, S. F. *Acta Crystallogr.* **1995**, *C51*, 18
24. Köhn, R. D.; Haufe, M.; Mihan, S.; Lilge, D. *J.Chem.Soc.,Chem. Commun.* **2000**, 1927
25. White, P. A.; Calabrese, J.; Theopold, K. H. *Organometallics* **1996**, *15*, 5473
26. Gibson, V. C.; Newton, C.; Redshaw, C.; Solan, G. A.; White, A. J. P.; Williams, D. J. *J. Chem. Soc., Dalton Trans.* **1999**, 827
27. Enders, M.; Fernandez, P.; Ludwig, G.; Pritzkow, H. *Organometallics* **2001**, *20*, 5005

Chapter 8

Olefin Copolymerization and Polymerization of Methyl Methacrylate Catalyzed by Tantalum(V)-Based Complexes

Wesley R. Mariott, Lauren M. Hayden, and Eugene Y.-X. Chen[*]

Department of Chemistry, Colorado State University,
Fort Collins, CO 80523–1872

Three different types of tantalum(V) complexes—tantalocene **1**, half-sandwich imido tantalum **2**, and bulky chelating diamide tantalum trimethyl **3**—have been investigated for ethylene–1-octene copolymerization and for polymerization of methyl methacrylate (MMA). **1** is inactive for the copolymerization upon activation with an excess of strong Lewis acids $M(C_6F_5)_3$ (M = B, Al), while **3** exhibits low copolymerization activity when activated with an aluminum imidzolide or a PMAO-IP activator. Complex **2**, when activated with the aluminum imidzolide activator, however, shows a high copolymerization activity of 1.2×10^6 g/(mol metal·atm·h) at a polymerization temperature of 140 °C, producing low density (0.898 g/mL) poly(ethylene-*co*-1-octene) copolymers having high molecular weight ($M_w = 127K$).

MMA polymerization in toluene by complex **1** upon activation with 2 equiv of $Al(C_6F_5)_3$ is sluggish; however, the same polymerization in *o*-dichlorobenzene is very rapid, producing quantitative yield of syndiotactic PMMA ([*rr*] = 74%) with a high molecular weight of $M_n = 1.3 \times 10^6$ Da in 5 min of polymerization time. A polymerization mechanism, which involves a chain initiation by the preformed binuclear aluminate anion and a bimetallic propagation via repeated intermolecular 1,4-Michael addition, is proposed.

Introduction

Unlike group 4 metallocene and related complexes, which are extensively used as structurally well-defined, typically highly active, single-site olefin polymerization catalysts (*1*), group 5 metal complexes, especially tantalum complexes, have not been widely studied and used as effective homogeneous catalysts for polymerization of vinyl monomers. Simple tantalocenes such as Cp_2TaCl_2, when mixed with an large excess of MAO, were reported to exhibit no ethylene polymerization activity (*2*). The half-sandwich tantalum(III) and niobium(III) dichloride or dialkyl complexes incorporating neutral diene ligands (i.e., isoelectronic structures of the group 4 metallocenes), however, are living ethylene polymerization catalysts and are also active for MMA polymerization upon activation with suitable activators (*2,3*). Other types of group 5 complexes that are recently reported to be active for ethylene polymerization include: tantalum complexes supported by the dianionic borollide (*4*) and tribenzylidenemethane (*5*) ligand, half-sandwich imido (*6*) and amidinate (*7*) compounds, tantalum aminopyridinato complexes (*8*), vanadium (*9*) and niobium (*10*) complexes incorporating hydridotris(pyrazolyl)borato ligands, as well as vanadium diimine/pyridine complexes (*11*).

The central objective of this work is to investigate the catalytic activity for ethylene–1-octene copolymerization and for polymerization of MMA, using three different types of tantalum(V) complexes: tantalocene **1**, half-tantalocene **2**, and non-tantalocene **3**.

1   **2**   **3**

Experimental Section

Methods and materials. All syntheses and manipulations of air-sensitive materials were carried out in flamed Schlenk-type glassware on a dual-manifold Schlenk line or in an argon-filled glovebox. NMR-scale reactions were conducted in Teflon-valve-sealed sample J-Young tubes. Solvents were first saturated with nitrogen and then dried by passage through activated alumina and Q-5[TM] catalyst prior to use. Deuterated hydrocarbon NMR solvents were dried over sodium/potassium alloy and distilled and/or filtered prior to use. MMA was degassed and dried over CaH_2 overnight, and then freshly vacuum-distilled before use. All feeds for olefin copolymerization were treated by passage through columns of alumina and Q-5™ catalyst prior to introduction into the reactor.

Tantalocene trimethyl **1** (*12*), half-sandwich imido tantalum dimethyl **2** (*13*), and bulky chelating diamide tantalum trimethyl **3** (*13*) were prepared according to the literature procedures, respectively. $B(C_6F_5)_3$ was obtained as a solid from Boulder Scientific Co. PMAO-IP was purchased from Akzo-Nobel, while all other reagents were purchased from Aldrich Chemical Co. $[HNMe(C_{18}H_{37})_2]^+[(C_6F_5)_3AlNC_3H_3N(Al(C_6F_5)_3)_3]^-$ (Al imidazolide) (*14*) was prepared from 2 equiv of $Al(C_6F_5)_3^{15}$ (*Extra caution should be exercised when handling this material due to its thermal and shock sensitivity.*) with 1 equiv each of imidazole and dioctadecylmethylamine in toluene. $Ph_3C^+[B(C_6F_5)_4]^-$ (*16*) and *rac*-(SBI)Zr(NMe$_2$)$_2$ (SBI = Me$_2$Si(Ind)$_2$) (*17*) were prepared according to the literature procedures, respectively.

Ethylene–1-octene copolymerization. A stirred 2 L Parr reactor was charged with about 740 g of mixed alkanes solvent (Isopar E, ExxonMobil Co.) and 118 g of 1-octene comonomer. Hydrogen was added as a molecular weight control agent by differential pressure expansion from a 75 mL addition tank at 300 psig. The reactor contents were heated to the polymerization temperature of 140 °C (or otherwise indicated) and saturated with ethylene at 500 psig. Catalysts and activators, as dilute solutions in toluene, were premixed in the glove box and transferred to a catalyst addition tank and injected into the reactor. The polymerization conditions were maintained for 15 min with ethylene added on demand. Heat was continuously removed from the reaction through an internal cooling coil. The resulting solution was removed from the reactor, quenched with isopropyl alcohol, and stabilized by an addition of 10 mL of a toluene solution containing approximately 67 mg of a hindered phenol antioxidant (Irganox™ 1010 from Ciba Geigy Corp.) and 133 mg of a phosphorus stabilizer (Irgafos™ 168 from Ciba Geigy Corp.). Between polymerization runs a wash cycle was conducted in which 850 g of Isopar-E was added to the reactor and the reactor was heated to 150 °C. The reactor was then emptied of the heated solvent immediately before beginning a new polymerization run.

Polymers were recovered by drying for 20 h in a vacuum oven set at 140 °C. High-temperature GPC analyses of polymer samples were carried out in 1,2,4-trichlorobenzene at 135 °C on a Waters 150C instrument. A polystyrene/polyethylene universal calibration was carried out using narrow molecular weight distribution polystyrene standards from Polymer Laboratories with 2,6-di-*tert*-butyl-4-methylphenol as the flow marker.

Methyl methacrylate polymerization. MMA polymerizations were performed in 50 mL Schlenk tubes with a septum and an external temperature-controlled bath on a Schlenk line or in an argon-filled glovebox. In a typical procedure, a tantalum complex (46.7 μmol) and an activator in a desired ratio (as indicated in Table 2) were loaded into the tube in the glovebox, and toluene (10 mL) or *o*-dichlorobenzene (DCB; 4 mL) was added. MMA (1.00 mL, 9.35 mmol) was added through the septum via a gastight syringe after stirring the catalyst and activator mixture for 10 min. The polymerization was quenched by adding 2 mL of acidified methanol after the measured time interval. The polymer product was precipitated into 50 mL of methanol, filtered, washed with methanol, and dried in a vacuum oven at 50 °C overnight to a constant weight. Glass transition temperatures of polymers were measured by differential scanning calorimetry (DSC 2920, TA Instruments, Inc.). Samples were first heated from room temperature to 180 °C. After being held at this temperature for 4 min, the samples were cooled to –20 °C at 10 °C/min and were then heated to 160 °C at 20 °C/min after being held at –20 °C for 4 min. GPC analyses of polymer samples were carried out at room temperature using THF as eluent on a Waters 150C instrument and calibrated using monodispersed polystyrene standards at a flow rate of 1.0 mL/min.

Results and Discussion

Ethylene–1-Octene Copolymerization

Table 1 summarizes the results from high-temperature ethylene–1-octene copolymerization studies using the tantalum(V) complexes **1–3** activated with various activators. Strong organo-Lewis acids $M(C_6F_5)_3$ (M = B, Al) can effectively abstract a methyl group from the tantalocene **1** to generate the corresponding cationic tantalocene species (*18*); however, it can be seen from the table that **1** has no olefin copolymerization activity even with an excess of Lewis acid activators (runs 1 and 2). Upon activation with an aluminum imidazolide activator, $[HNMe(C_{18}H_{37})_2]^+[(C_6F_5)_3AlNC_3H_3N(Al(C_6F_5)_3]^-$, the half-sandwich imido tantalum dimethyl **2** is very active; the activated catalyst produced 50 g of a copolymer within 15 min using 5.0 μmol of **2**, giving an activity of 1.2×10^6 g/(mol·atm·h) (run 3). This copolymer has a low density of 0.898, indicative of a large 1-octene incorporation in the copolymer. More impressively, this catalyst produces the copolymer with a high molecular weight of 127K, which is ~64% higher than the copolymer produced by a "constrained geometry" titanium catalyst (*13*), under otherwise similar polymerization conditions. The latter titanium catalyst, however, exhibits higher copolymerization activity.

The activated complex **2** is still very active at a polymerization temperature of 160 °C, giving an activity of 5.4×10^5 g/(mol·atm·h) (run 4). Despite a reduced polymerization activity at this temperature, the copolymer still has a low density of 0.900, a high molecular weight of 109K, and a molecular weight distribution of 2.20. Other activators such as PMAO-IP give lower polymerization activity, producing the copolymer with higher density but still having a high molecular weight of 125K (run 5).

The non-metallocene, bulky chelating diamide tantalum trimethyl **3** is active but exhibits low polymerization activity toward copolymerization of ethylene and 1-octene upon activation with the aluminum imidazolide or PMAO-IP activator (runs 6 and 7). The aminopyridinato tantalum complexes (**4**, **5**), however, are reported to be highly active, for homopolymerization of ethylene upon activation with MAO (*8*).

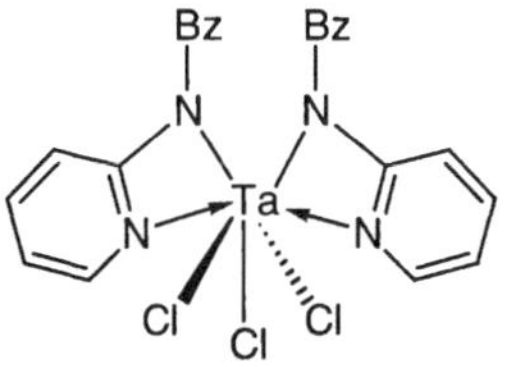

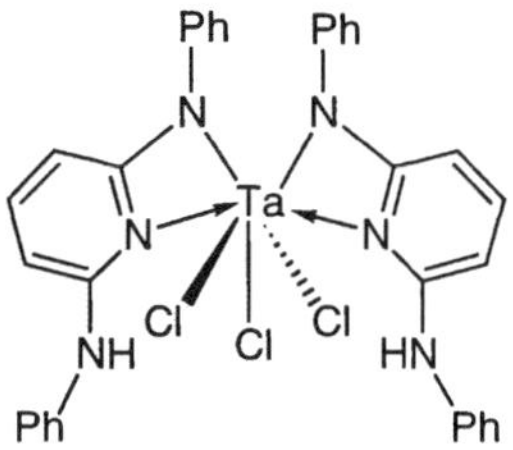

Table 1. Ethylene and 1-Octene Copolymerization Results with Ta(V) Complexes [a]

no	cat	co-cat	Cat/ Cocat (μmol)	T_p (°C)	yield (g)	A [b] (10^4)	density (g/mL)	M_w [c] (10^5)	PDI
1	1	borane	1.0 /4.0	140	trace	0			
2	1	alane	1.0 /4.0	140	trace	0			
3	2	Al [d]	5.0 /5.0	140	50.2	121	0.898	1.27	2.29
4	2	Al [d]	5.0 /5.0	160	22.4	54	0.900	1.09	2.20
5	2	PMAO -IP	20 /10000	130	14.8	9	0.917	1.25	2.72
6	3	Al [d]	2.0 /2.0	140	3.0	18	0.900		
7	3	PMAO -IP	20 /10000	140	1.9	1			

[a] Conditions: 118 g octene, 5 mmol of H_2, 500 psig ethylene, 740 g isopar-E, and 15 min. [b] Activity (A) in units of g/(mol·atm·h). [c] GPC relative to polystyrene standards. [d] Al imidazolide: $[HNMe(C_{18}H_{37})_2]^+[(C_6F_5)_3AlNC_3H_3N(Al(C_6F_5)_3)]^-$.

Methyl Methacrylate Polymerization

The non-tantalocene **3** exhibits no MMA polymerization activity upon activation with various activators, whereas the polymerization behavior of the half-sandwich imido tantalum complex **2** will be published elsewhere. Upon activation with 1 equiv of $B(C_6F_5)_3$ or $Al(C_6F_5)_3$ in toluene, the tantalocene **1** did not produce any solid polymer in the MMA polymerizations at ambient temperature for 12 h. Activation with 2 equiv of $B(C_6F_5)_3$ yielded the same result; however, active polymerization systems were generated when 2 equiv of $Al(C_6F_5)_3$ was used for the activation, regardless of addition sequence. Thus, the activated catalyst polymerization by first mixing Cp_2TaMe_3 with 2 equiv of $Al(C_6F_5)_3$ before adding MMA produced PMMA in quantitative yield at a MMA/initiator ratio of 200 for 10 h (Table 2, entry 4). A similar activity is observed via activated monomer polymerization by first mixing MMA with 2 equiv of $Al(C_6F_5)_3$ before adding Cp_2TaMe_3 (entry 5). These findings seem to be consistent with the observations that the reactions of Cp_2TaMe_3 with 1 equiv of $Al(C_6F_5)_3$ or with 1 or 2 equiv of $B(C_6F_5)_3$ in toluene produce a mixture of species, while the reaction with 2 equiv of $Al(C_6F_5)_3$ in either toluene or bromobenzene cleanly generates a tantalocene cation paired with an μ–methyl bridged binuclear aluminate anion (i.e., $Cp_2TaMe_2^+[(C_6F_5)_3Al\text{-}CH_3\text{-}Al(C_6F_5)_3]^-$; Scheme 1) (*13,18*).

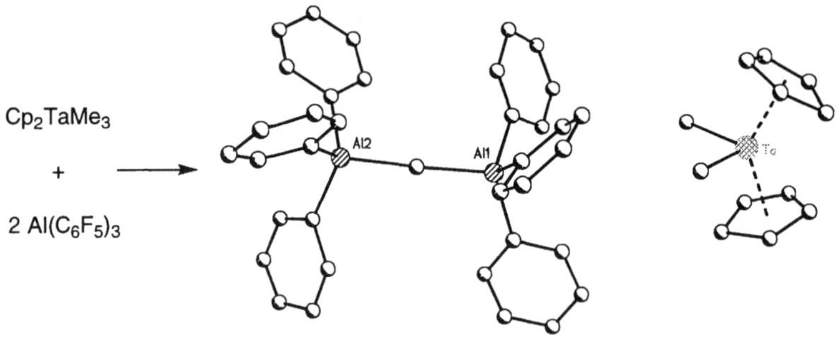

Scheme 1

Number average molecular weights of the PMMA produced from **1** in toluene are significantly higher than the calculated values, and molecular weight distributions are also broad (PDI from 3.22 to 5.42). Initiator efficiencies are low (from 5.6% to 12.6%), whereas the syndiotacticity of PMMA is appreciable with [*rr*] triads ranging from 73 to 76%.

Solubility of the activated form of the catalyst (i.e., $Cp_2TaMe_2^+[(C_6F_5)_3Al$-$CH_3$-$Al(C_6F_5)_3]^-$), derived from the reaction of **1** with 2 equiv of $Al(C_6F_5)_3$, is limited in toluene; hence, the heterogeneity during the course of the MMA polymerization is believed to cause the low catalysts efficiency and broad polymer molecular weight distribution. To address this solubility issue and to investigate the solvent polarity effect on polymerization, MMA polymerizations were carried out in o-dichlorobenzene (Table 2, entries 6 to 10). Indeed, tantalocene **1**, when activated with 2 equiv of $Al(C_6F_5)_3$, becomes highly active for MMA polymerization in o-dichlorobenzene. Thus, PMMA was produced in quantitative yield in just 5 min (entry 6), reflecting a 120-fold increase in polymerization activity as compared to the polymerization carried out in toluene. The molecular weight distribution was also narrower, and more interestingly, the polymer has much higher molecular weight ($M_n = 1.3 \times 10^6$ Da), whereas the syndiotacticity is kept almost constant ($[rr] = 74\%$). The repeated polymerization on a high-vacuum line yielded a similar polymerization activity, yield, and syndiotacticity (entry 7). The polymerization by **1** in combination with 1 equiv of $Al(C_6F_5)_3$ produced PMMA in just 3% yield, even in an much extended polymerization time of 2 h, whereas **1**, when activated with 1 or 2 equiv of $B(C_6F_5)_3$ or just 1 equiv of $Ph_3CB(C_6F_5)_4$, did not yield any solid polymer in 2 h (entries 9 and 10).

Table 2. MMA Polymerization Results with Cp_2TaMe_3 [a]

no	Activator	solvent	t_p (min)	yield (%)	$[rr]$ [b] (%)	T_g (^{o}C)	M_n [c] (10^5)	M_w/M_n
1	$B(C_6F_5)_3$	T	720	0				
2	$Al(C_6F_5)_3$	T	720	0				
3	$2B(C_6F_5)_3$	T	720	0				
4	$2Al(C_6F_5)_3$	T	600	100	73	121	1.59	5.42
5[d]	$2Al(C_6F_5)_3$	T	600	100	76		3.47	3.22
6	$2Al(C_6F_5)_3$	DCB	5	100	74	130	13.0	2.98
7[e]	$2Al(C_6F_5)_3$	DCB	5	95	74	124		
8	$Al(C_6F_5)_3$	DCB	120	3	72			
9	$2B(C_6F_5)_3$	DCB	120	0				
10	$Ph_3CB(C_6F_5)_4$	DCB	120	0				

[a] Conditions: 46.7 μmol initiator (I); mole ratio MMA/I = 200; 10 mL toluene (T) or 4 mL o-dichlorobenzene (DCB); 25 °C. Catalyst and activator are mixed in situ with a desired ratio as indicated in the complex column. [b] Determined by ^{1}H NMR spectroscopy. [c] GPC relative to polystyrene standards. [d] For run 5, MMA monomer was first mixed with $Al(C_6F_5)_3$ followed by addition of Cp_2TaMe_3. [e] Carried out on a high vacuum line.

The above observations clearly rule out the possibility of a direct initiation by the tantalocene cation ($Cp_2TaMe_2^+$) because all three activators (i.e., $B(C_6F_5)_3$, $Al(C_6F_5)_3$ and $Ph_3CB(C_6F_5)_4$) generate the same cation $Cp_2TaMe_2^+$. Complex **1** is active, however, only when 2 equiv of $Al(C_6F_5)_3$ is used, the reaction of which cleanly produces $Cp_2TaMe_2^+[(C_6F_5)_3Al\text{-}CH_3\text{-}Al(C_6F_5)_3]^-$ (*vide supra*). The crucial difference is, therefore, the formation of the binuclear aluminate anion in the active MMA polymerization system. A proposed polymerization initiated by this binuclear anion is outlined in Scheme 2. In this proposed mechanism, monomer activation proceeds with alane transfer from the binuclear anion to MMA, followed by chain initiation via nucleophilic attack of the alane-activated MMA by methide group of the methyl aluminate anion, which itself is not capable of initiating the inactivated MMA. The resulting enolaluminate undergoes intermolecular 1,4-Michael addition to the activated MMA, providing a repeated reaction sequence for a bimetallic chain propagation. The molecular structure of the alane-activated MMA and formation of the enolaluminate propagating species in the MMA polymerization initiated by group 4 metallocenium aluminates have been reported earlier (*19*).

The MMA polymerization was inhibited by the presence of a free radical scavenger, galvinoxyl, but it was found out that galvinoxyl rapidly decomposes

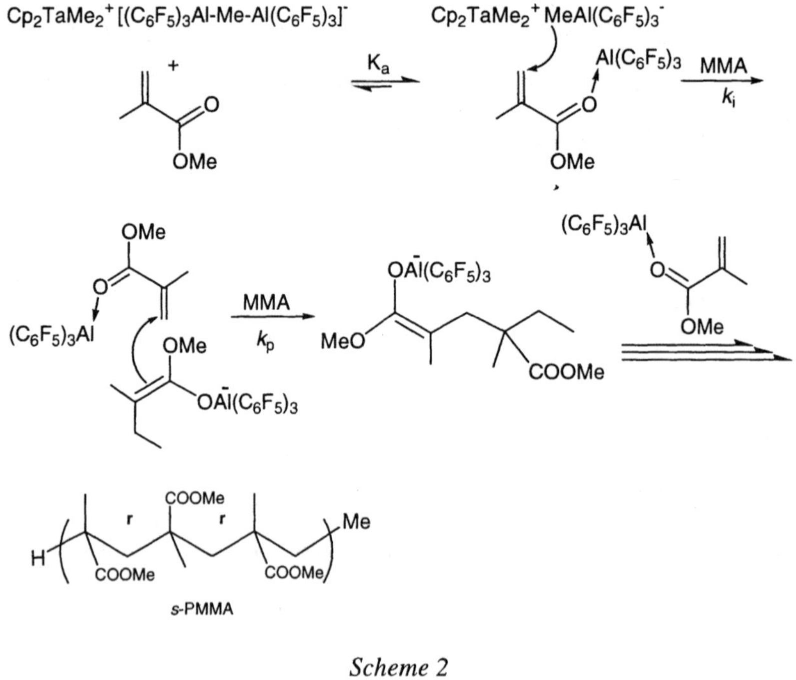

Scheme 2

Al(C$_6$F$_5$)$_3$. Nevertheless, the initiator (**1** + 2 Al(C$_6$F$_5$)$_3$) in *o*-dichorobenzene does not polymerize vinyl acetate, a monomer that can be polymerized radically but not anionically. Furthermore, control runs with **1** or Al(C$_6$F$_5$)$_3$ alone in *o*-dichorobenzene were completely inactive for MMA polymerization. These experiments rule out a radical initiation due to possible reaction of the alane with *o*-dichorobenzene. NMR experiments also show neither reaction nor decomposition of the alane in *o*-dichorobenzene at room temperature or 70 °C for 16 h.

To provide additional evidence for the proposed bimetallic mechanism initiated by the binuclear anion, MMA polymerization was carried out at room temperature in *o*-dichlorobenzene using *rac*-(SBI)Zr(NMe$_2$)$_2$ + 2Al(C$_6$F$_5$)$_3$, the reaction of which produces *rac*-(SBI)Zr(NMe$_2$)$^+$[(C$_6$F$_5$)$_3$Al-NMe$_2$-Al(C$_6$F$_5$)$_3$]$^-$ (Scheme 3). The polymerization at a [MMA]/[initiator] ratio of 200 also produced syndiotactic PMMA ([*rr*] = 71%) in quantitative yield within 2 h, and again, activations with 1 or 2 equiv of B(C$_6$F$_5$)$_3$ or just 1 equiv of Al(C$_6$F$_5$)$_3$ did not yield any solid polymer.

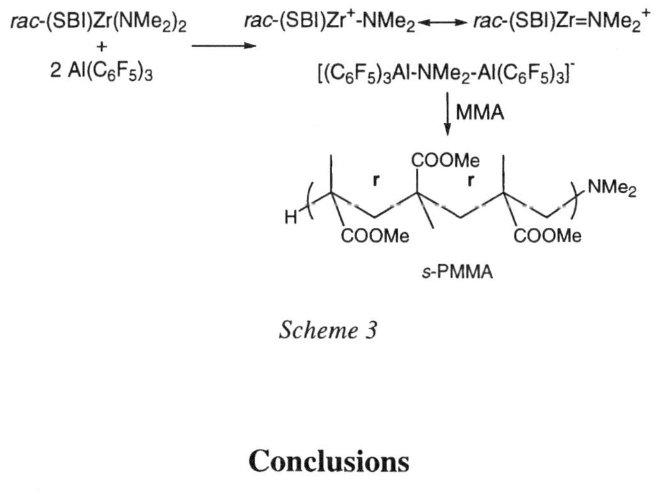

Scheme 3

Conclusions

Among the three types of the tantalum(V) complexes investigated for ethylene–1-octene copolymerization, the half-sandwich imido tantalum complex **2** not only shows high activity, but also produces high molecular weight copolymer with a large amount of bulky olefin incorporation. This type of group 5 complexes warrants further investigation and development. In MMA polymerizations, the initiator composed of the tantalocene **1** in combination with 2 equiv of Al(C$_6$F$_5$)$_3$ is extremely active for MMA polymerization in polar solvents such as *o*-dichlorobenzene, producing syndiotactic PMMA with ultrahigh molecular weight. The proposed polymerization mechanism suggests that a variety of binuclear aluminate anions having a general formula of $[(C_6F_5)_3Al\text{-}\mathbf{X}\text{-}Al(C_6F_5)_3]^-$, where $\mathbf{X}$ is an alkyl R, hydrido H, amido R_2N, or cyanide CN, etc., can be very effective initiators for MMA polymerization.

Acknowledgments. The authors thank Gordon R. Roof of The Dow Chemical Company for carrying out high temperature ethylene–1-octene copolymerizations. We also thank Colorado State University, the donors of the Petroleum Research Fund, administrated by the American Chemical Society, and the NSF-REU program, for financial support.

References

1   For recent reviews, see: (a) Coates, G. W.; Hustad, P. D.; Reinartz, S. *Angew. Chem., Int. Ed. Engl.* **2002**, *41*, 2236–2257. (b) Erker, G. *Acc. Chem. Res.* **2001**, *34*, 309–317. (c) Chum, P. S.; Kruper, W. J.; Guest, M. J. *Adv. Mater.* **2000**, *12*, 1759–1767. (d) Gladysz, J. A., Ed. *Chem. Rev.* **2000**, *100*, 1167–1682. (e) Marks, T. J.; Stevens, J. C., Eds. *Top. Catal.* **1999**, *7*, 1–208. (f) Britovsek, G. J. P.; Gibson, V. C.; Wass, D. F. *Angew. Chem., Int. Ed. Engl.* **1999**, *38*, 428–447. (g) Jordan, R. F., Ed. *J. Mol. Catal.* **1998**, *128*, 1–337. (h) McKnight, A. L.; Waymouth, R. M. *Chem. Rev.* **1998**, *98*, 2587–2598. (i) Piers, W. E. *Chem. Eur. J.* **1998**, *4*, 13–18. (j) Kaminsky, W.; Arndt, M. *Adv. Polym. Sci.* **1997**, *127*, 144–187. (k) Bochmann, M. *J. Chem. Soc., Dalton Trans.* **1996**, 255–270. (l) Brintzinger, H.-H.; Fischer, D.; Mülhaupt, R.; Rieger, B.; Waymouth, R. M. *Angew. Chem., Int. Ed. Engl.* **1995**, *34*, 1143–1170.

2   Mashima, K.; Fujikawa, S.; Nakamura, A. *J. Am. Chem. Soc.* **1993**, *115*, 10990–10991.

3   (a) Matsuo, Y.; Mashima, K.; Tani, K. *Angew. Chem., Int. Ed. Engl.* **2001**, *40*, 960–962. (b) Mashima, K. *Macromol. Symp.* **2000**, *159*, 69–76. (c) Mashima, K.; Fujikawa, S.; Tanaka, Y.; Urata, H.; Oshiki, T.; Tanaka, E.; Nakamura, A. *Organometallics* **1995**, *14*, 2633–2640. (d) Mashima, K.; Fujikawa, S.; Urata, H.; Tanaka, E.; Nakamura, A. *J. Chem. Soc. Chem. Comm.* **1994**, 1623–1624.

4   Bazan, G. C.; Donnelly, S. J.; Rodriguez, G. *J. Am. Chem. Soc.* **1995**, *117*, 2671–2672.

5   Rodriguez, G.; Bazan, G. C. *J. Am. Chem. Soc.* **1995**, *117*, 10155–10156.

6   (a) Coles, M. P.; Dalby, C. I.; Gibson, V. C.; Little, I. R.; Marshall, E. L.; da Costa, M. H. R.; Mastroianni, S. *J. Organomet. Chem.* **1999**, *591*, 78–87. (b) Witte, P. T.; Meetsma, A.; Hessen, B. *Organometallics* **1999**, *18*, 2944–2946. (c) Antonelli, D. M.; Leins, A.; Stryker, J. M. *Organometallics* **1997**,

16, 2500–2502. (d) Coles, M. P.; Gibson, V. C. *Polym. Bull.* **1994**, *33*, 529–533.

7   Decker, J. M.; Geib, S. J.; Meyer, T. Y. *Organometallics* **1999**, *18*, 4417–4420.

8   Hakala, K.; Löfgren, B.; Polamo, M.; Leskelä, M. *Macromol. Rapid Commun.* **1997**, *18*, 635–638.

9   Scheuer, S.; Fischer, J.; Kress, J. *Organometallics* **1995**, *14*, 2627–2629.

10 Jaffart, J.; Nayral, C.; Choukroun, R.; Mathieu, R.; Etienne, M. *Eur. J. Inorg. Chem.* **1998**, 425–428.

11 Reardon, D.; Conan, F.; Gambarottta, S.; Yap, G.; Wang, Q. *J. Am. Chem. Soc.* **1999**, *121*, 9318–9325.

12 Schrock, R. R.; Sharp, P. R. *J. Am. Chem. Soc.* **1978**, *100*, 2389–2399.

13 Feng, S.; Roof, G. R.; Chen, E. Y.-X. *Organometallics* **2002**, *21*, 832–839.

14 (a) LaPointe, R. E.; Roof, G. R.; Abboud, K. A.; Klosin, J. *J. Am. Chem. Soc.* **2000**, *122*, 9560–9561. (b) LaPointe, R. E. PCT Int. Appl. WO 9942467, 1999.

15 Biagini, P.; Lugli, G.; Abis, L.; Andreussi, P. U.S. Pat. 5,602269, 1997.

16 (a) Chien, J. C. W.; Tsai, W.-M.; Rausch, M. D. *J. Am. Chem. Soc.* **1991**, *113*, 8570–8571. (b) Ewen, J. A.; Elder, M. J. Eur. Pat. Appl. EP 0,426,637, **1991**.

17 Christopher, J. N.; Diamond, G. M.; Jordan, R. F.; Petersen, J. L. *Organometallics* **1996**, *15*, 4038–4044.

18 Chen, E. Y.-X.; Abboud, K. A. *Organometallics* **2000**, *19*, 5541–5543.

19 Bolig, A. D.; Chen, E. Y. X. *J. Am. Chem. Soc.* **2001**, *123*, 7943–7944.

Catalysts from Groups 8–10 Metal Complexes

The Alternating Copolymerization of Epoxides and Carbon Monoxide: A Novel Way to Polyesters

Markus Allmendinger[1], Robert Eberhardt[1], Gerrit A. Luinstra[2], Ferenc Molnar[2], and Bernhard Rieger[1,*]

[1]Department of Materials and Catalysis, University of Ulm, Albert-Einstein-Allee 11, 89069 Ulm, Germany
[2]BASF Aktiengesellschaft GKT/P B1, 87056 Ludwigshafen, Germany

The background of investigations on polyhydroxyalkanoates is discussed in terms of biological and synthetic approaches. Special emphasis is placed on an overview of the newly discovered direct alternating copolymerization reaction of epoxides and carbon monoxide. It is demonstrated that polyhydroxyalkanoates, like PHB, are available by this new type of copolymerization reaction. Further insight into mechanistic aspects is discussed.

Background of PHB research

Durability and resistance to all forms of degradation, combined with special performance characteristics, were expected to be the major advantage of many synthetic polymers like for example polypropylene. However, these valuable properties cause a disposal problem due to the durability and undefined environmental fate of most of currently used polymer materials. For this reason waste management of polymers is of high interest and biodegradable representatives are likely to be favored in short term applications, e.g. packaging, agricultural films or medical use. In addition, not only high molecular weight products may have a broad market, but also biodegradable oligomers, which are currently investigated as plasticizers, drug delivery systems or macromolecular building blocks (*1*).

Biochemical approach

One of the most promising families of biocompatible and biodegradable polymers are polyhydroxyalkanoates (PHAs) (*2*). These were already discovered in 1925 by Lemoigne in natural bacterial sources (Figure 1) (*3*). Later it was found that certain bacteria like for example *Alcaligenes eutrophus* use these polymers as storage materials comparable to the role of starch in plants (*4*).

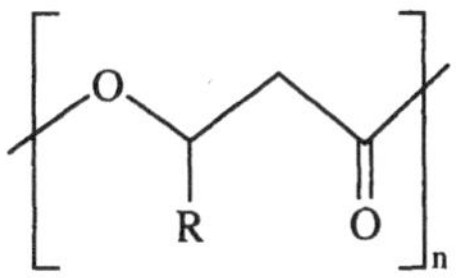

Figure 1. Structure of polyhydroxyalkanoates; R = alkyl, haloalkyl, aryl etc.

The most prominent and widespread member of this group of biopolymers is polyhydroxybutyrate (PHB), which can be found in nature only as a strictly isotactic material of *R* configuration (*R*-PHB). This structural limitation leads to a brittle, high-molecular weight polymer with the major drawback that its decomposition already starts at melt processing (*5*). To overcome these restrictions several attempts were made to incorporate comonomers into the PHB-polyester chain. One successful example is a copolymer of 3-hydroxybutyrate and 3-hydroxyvalerate (PHB/V), which was produced as BIOPOL® in a multi-ton scale by ICI (*6*). This polymer has physical properties similar to those of *i*-polypropylene; it can be melt-processed with reduced risk of thermal degradation and is still biodegradable (*7*). The commercialization was much less successful, as the polymer properties would suggest. High manufacturing costs and difficulties to control molecular weight, functionality, and consequently, to fine-tune polymer properties were the major problems of the enzymatic production process.

116

Nonbiochemical routes

An alternative, nonbiochemical route to PHB uses the (metal) catalyzed ring opening polymerization of β-butyrolactone (β-BL) (*8*). The lactone can thereby for example be obtained by a carbonylation reaction of epoxides (Figure 2).

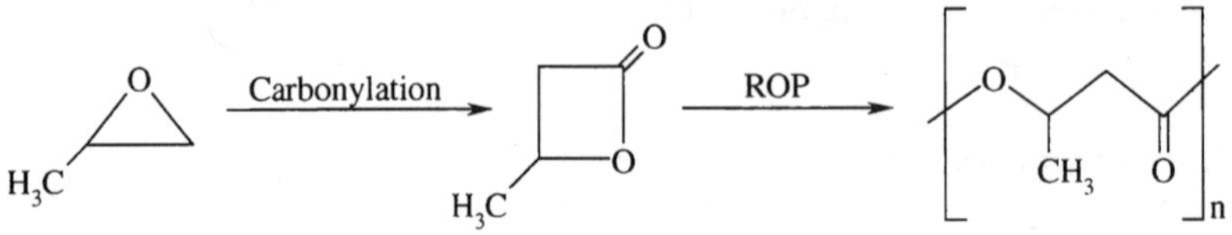

Figure 2. Carbonylation of propylene oxide and subsequent ring opening polymerization of β-butyrolactone to PHB

Despite of a few promising reports on the lactone synthesis (*9*) and ongoing investigations to control polymer chain stereochemistry (*10*), this synthetic strategy cannot yet be considered a feasible method for an industrial production of PHB. The drawbacks here are the low reactivity of the monomer, difficulties to control the stereochemistry and in consequence the expensive multi-step synthesis of enantiomerically pure β-BL.

A new, promising synthetic approach to polyalkanoates, especially to PHB, is the direct alternating copolymerization of epoxides and carbon monoxide.

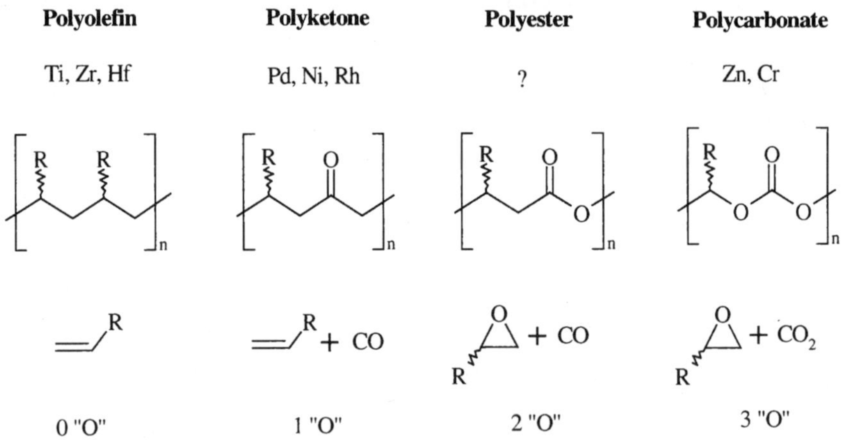

Polyolefin	Polyketone	Polyester	Polycarbonate
Ti, Zr, Hf	Pd, Ni, Rh	?	Zn, Cr
0 "O"	1 "O"	2 "O"	3 "O"

Figure 3. Series of known polymers from technical monomers – aliphatic polyesters as a missing link

The catalytic synthesis of polyolefins (*11*), polyketones (*12*) and polycarbonates (*13*) through insertion reactions of small, technically available, cheap monomers is established practice in industrial production. A comparable process leading directly to aliphatic polyesters was unprecedented (Figure 3). This is even more surprising since carbonylation reactions of epoxides to yield low molecular weight products (lactones, β-hydroxyester) are known for a long time (*9,14*).

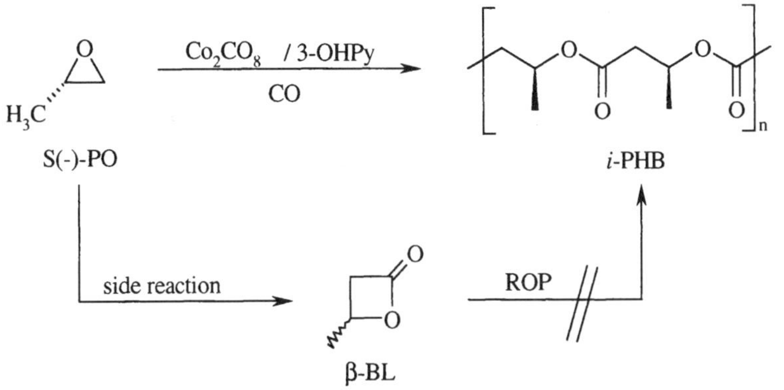

Figure 4. Alternating copolymerization of epoxides and carbon monoxide

Recently, we reported on the first alternating copolymerization reaction of epoxides and carbon monoxide, which was proven not to proceed via a lactone intermediate (Figure 4) (*15*). Here we present a summary of our studies related to the cobalt catalyzed alternating copolymerization of propylene oxide (PO) and carbon monoxide, affording atactic and isotactic polyhydroxybutyrates depending on the stereochemistry of the starting epoxide.

Alternating copolymerization of epoxides and carbon monoxide

A number of reports on the carbonylation reaction of epoxides were published when we started our research on the synthesis of aliphatic polyesters from epoxides and CO (*9,14*). One of the first mentioning the formation of oligomeric products was a paper by Furukawa et al. in 1965, using a mixture of Co(acac)$_3$ and Et$_3$Al as catalyst (*16*). A more selective catalysis was patented in 1994 by Drent, using a mixture of Co$_2$CO$_8$/3-hydroxypyridine (3OH-Py), primarily for the formation of lactones from epoxides and CO (*17*). It was much

118

to our surprise, that the same catalytic mixture gave polyhydroxybutyrates in our hands, a finding that prompted us to investigate this reaction in more detail.

Co$_2$CO$_8$ / 3OH-Py

We decided to use an autoclave that was equipped with an ATR-IR spectrometer to allow online monitoring of lactone and polyester formation and at the same time obtain information on catalytically active organometallic species (*18*).

Figure 5 depicts the concentration dependence of the formation of lactone and polyester from PO and CO as function of time. To a first approximation the lactone (1829 cm^{-1}) concentration develops linearly with time, and clearly remains the minor product. The rate of formation of the major product, polyhydroxybutyrate (1744 cm^{-1}), is independent of the presence of lactone and increases steadily during the entire reaction time of about 17 h. This fits well with the concentration of propylene oxide, which decreases over the same time interval until complete consumption. 1H-NMR analysis of the reaction mixture revealed in addition that the amount of β-butyrolactone is not larger than 15% at any stage of the reaction profile.

We repeated the reaction under the same conditions without epoxide but with butyrolactone as reagent to investigate whether the polymerization occurs via the lactone as intermediate. No polymerization product could be identified in phase A (Figure 6), showing that lactone does not undergo a ring opening polymerization in the presence of the catalyst mixture. Polyester formation though starts immediately after addition of epoxide to the reaction mixture (B) (*19*) and proceeds (C) until all epoxide is consumed. The lactone concentration does not decrease but even grows marginally during the process. This shows clearly that the polyester is directly produced from propylene oxide and CO, and that the lactone is a side product and not an intermediate in PHB-formation.

Further proof for a direct copolymerization reaction originates from NMR analysis of the polymer stereoregularity. The latter depends on the chirality of the propylene oxide starting material. Polymerizations using enantiomerically pure S-(-)PO or R-(+)PO give exclusively isotactic PHB (Table 1, entry #4). We isolated the same isotactic material by repeating the copolymerization of enantiomerically pure PO in the presence of racemic β-butyrolactone (Table 1, entry #5). This observation shows that lactone is not incorporated in the polymer chain, which should have led to a reduction of the PHB isotacticity, and constitutes an independent indication for a direct polymer formation from PO and CO. It is further fully consistent with a direct regiospecific alternating copolymerization of epoxide and carbon monoxide monomers.

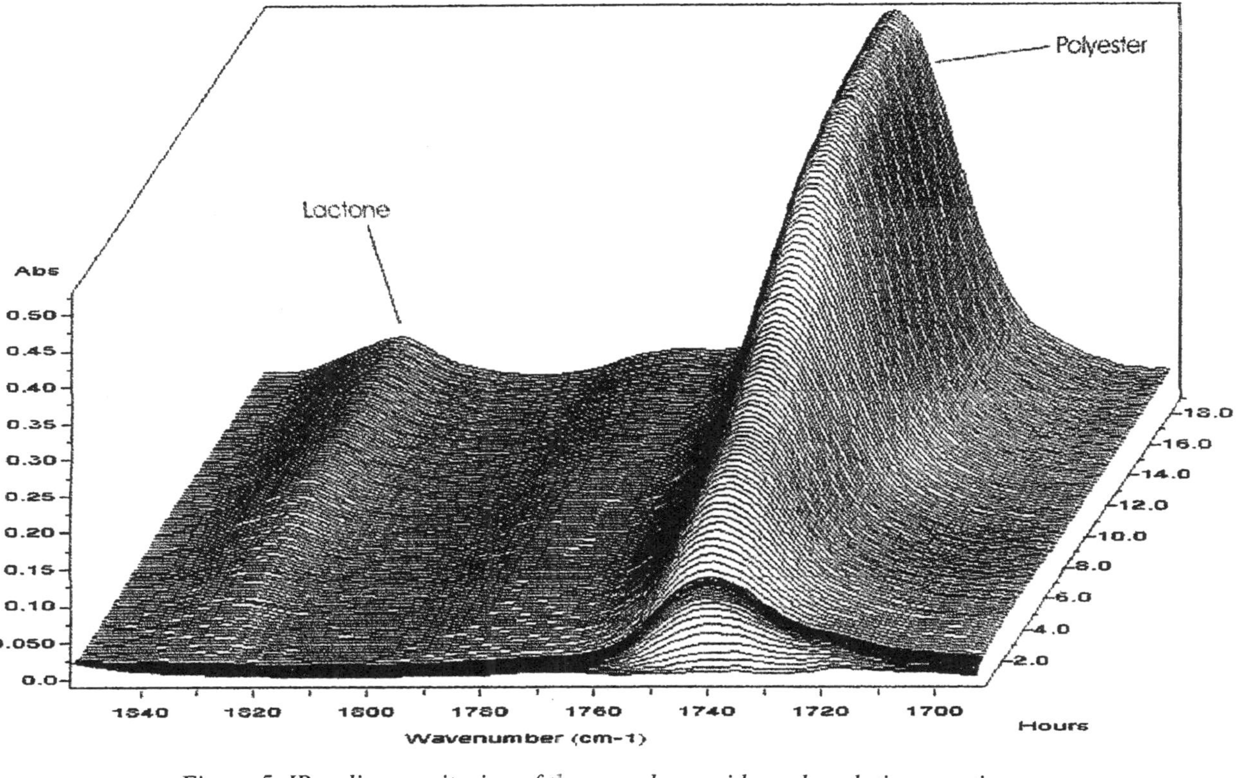

Figure 5. IR-online monitoring of the propylene oxide carbonylation reaction (catalyst: Co_2CO_8/3-hydroxypyridine)

(Reproduced from reference 15. Copyright 2002 American Chemical Society.)

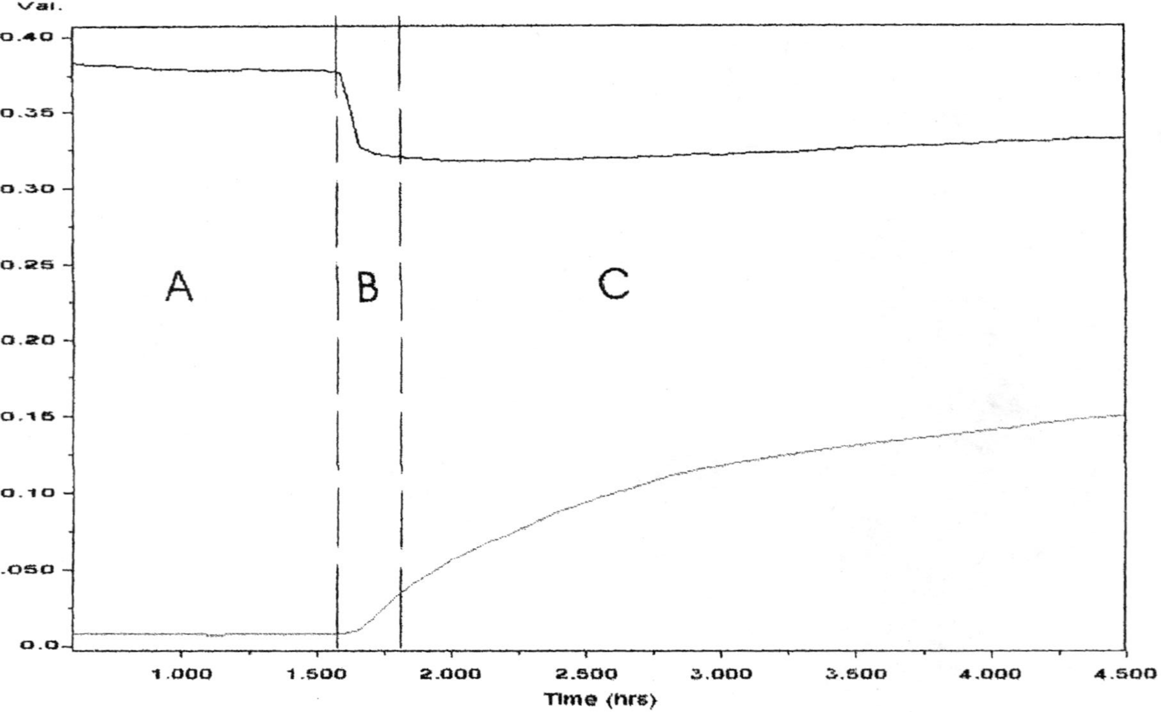

Figure 6. Profile of polyester (red) and lactone (black) from IR-online monitoring of reaction mixture with lactone added from the beginning and propylene oxide added after 1.5 hours (catalyst: $Co_2(CO)_8$/3-hydroxypyridine)

(Reproduced from reference 15. Copyright 2002 American Chemical Society.)

Table 1. Copolymerizations of epoxides with carbon monoxide[a]

No	Monomer	Convers. of epoxide[b]	Yield Polyester [g][c]	M_w [g/mol] / D[d]
1	EO / 5 ml	80%	2.8[h]	7400 / 1.6
2a	EO / 20 ml[f]	25%	7.4	5100/ 1.7
b	EO / 20 ml[g]	30%	8.8	10,500 /1.8
3	rac-PO / 2mL	>90%	1.8 (atactic)[e]	3800 / 2.0
4	S(-)-PO / 2ml	>90%	1.4 (isotactic)[e] [i]	4200 / 1.3
5	S(-)-PO 2ml/ rac-BL 0.5 ml	>90%	1.2 (isotactic)[e]	4100 / 1.4

[a] diglyme (10ml), $Co_2(CO)_8$ 109 mg, 3-hydroxypyridine 61 mg, 4h. (1), 18h. (3,4,5); CO 6o bar, 75°C [b] conversion determined by 1H-NMR of the reaction mixture. [c] Yield after precipitation in methanol (1,2,4,5) or hexane/diethyl ether (3). [d] molecular weight data were obtained by gel permeation chromatography (GPC) using a Waters 150C ALC/GPC in $HCCl_3$, relative to narrow polystyrene standards. [e] tacticity determined by 1H- and 13C-NMR (*20*). [f] Diglyme 80 mL, 3-hydroxypyridine 760 mg, $Co_2(CO)_8$ 684 mg, CO 60 bar, 80°C, 1h. [g] as in [f] and dimethoxypropane 1.1 g. [h] melting point of polyhydroxypropionate is 62°C determined by DSC. [i] melting point of i-PHB is 129°C determined by DSC.

Mechanistic considerations

A possible mechanism pertaining the formation of regio- and stereoregular polyhydroxybutyrates is based on a sequence of alternating ring opening reactions of epoxide and CO-insertion reactions at a tetracarbonyl cobaltate anion. N-Donors, such as pyridine compounds, are known to disproportionate the Co-Co bond in carbonyl complexes leading to the formation of the $Co(CO)_4^-$ -anion and Co(II) fragments (*21*). In accordance, only a few minutes after the addition of 3-hydroxypyridine, most of the Co_2CO_8 is converted into ion pairs. The characteristic strong carbonyl absorption ($1889cm^{-1}$) of $Co(CO)_4^-$ dominates the IR-spectrum (Figure 7). Experiments to induce carbonylation reactions by using $Co(CO)_4^-$ salts with cations of low Lewis acidity (e.g. $PPNCo(CO)_4$) failed. Also, unsubstituted pyridine in combination with Co_2CO_8 give negligible PO conversion. These observations suggest, that the pyridine substituent plays a crucial role in catalyst formation, when starting from $Co_2(CO)_8$. Systematic variation of the pyridine substitution (3-hydroxypyridine, nicotinic acid and nicotinic acid amide) shows that besides induction of the dicobaltoctacarbonyl disproportionation reaction, the acidic 3-hydroxy unit plays an important role, most probably to polarize the epoxide monomer (*22,23*).

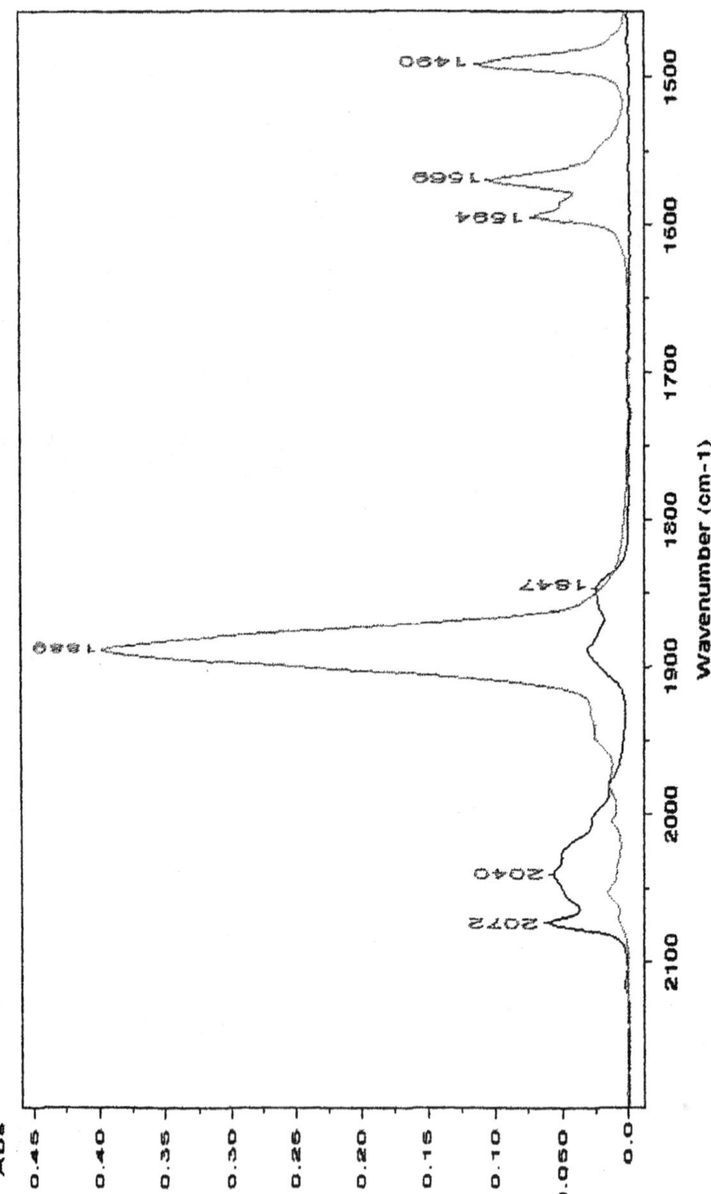

Figure 7. IR spectrum of Co_2CO_8 (2072, 2040 and 1847 cm^{-1}) in diglyme before (black) and after (red, $Co(CO)_4^-$ 1889cm^{-1}) addition of 3-hydroxypyridine

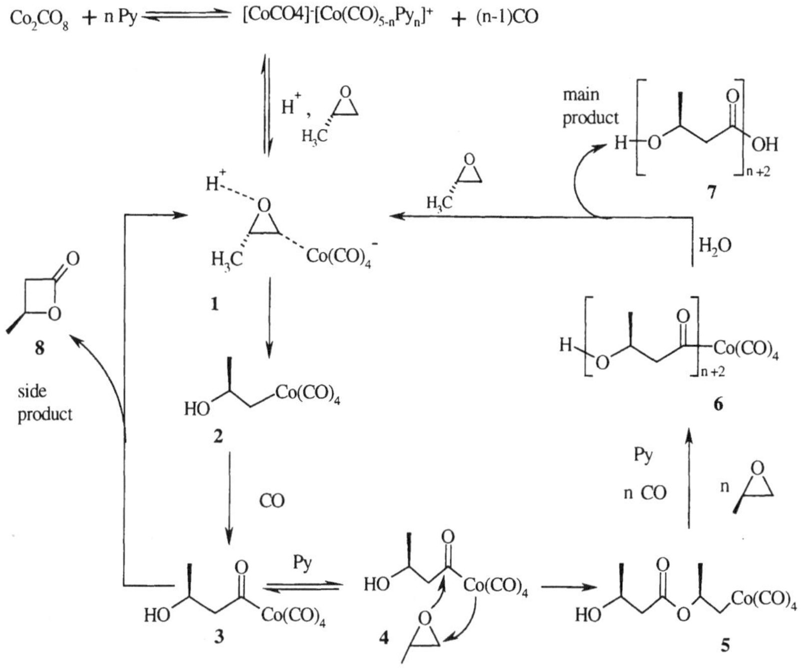

$$Co_2CO_8 + n\,Py \rightleftharpoons [CoCO4]^-[Co(CO)_{5-n}Py_n]^+ + (n-1)CO$$

Figure 8. Proposed reaction mechanism

We suggest that the OH-group of 3-hydroxypyridine (or of Co^{2+}-coordinated 3-hydroxypyridine) delivers a proton, that together with a $Co(CO)_4^-$-anion induces the first ring opening reaction of an epoxide molecule. This corresponds to the observations of Heck and others, who showed that $HCo(CO)_4$ reacts smoothly with epoxides even at low temperatures to give compound 2 (Figure 8). This reactive species undergoes a facile carbonylation in the presence of carbon monoxide to form the acyl complex 3 (*24*). Its characteristic absorptions (2107, 2043, 2022, 2005, 1717 cm^{-1}, Figure 9) were actually detected in the copolymerization reaction of propylene oxide/CO with $Co_2(CO)_8/3HO$-Py as catalyst.

Heck found in his pioneering study no evidence for multiple ring opening and insertion reactions of epoxides and CO. He reported that compound 3 affords methyl 3-hydroxybutyrate after treatment with methanol (*24*). These findings distinguish our copolymerization reaction from the former carbonylation and leave us with the question of the epoxide insertion into the cobalt-acyl bond ($3\rightarrow4\rightarrow5$, Figure 8).

To gain insight into this reaction step, we used a defined catalyst system. We now started from the easily accessible $Et_4NCo(CO)_4$ dissolved in a solution of propylene oxide and diglyme. The presence of pure $Co(CO)_4^-$ (1889 cm^{-1}) was confirmed by ATR-IT spectroscopy (Figure 10, black). Addition of HBF_4 under

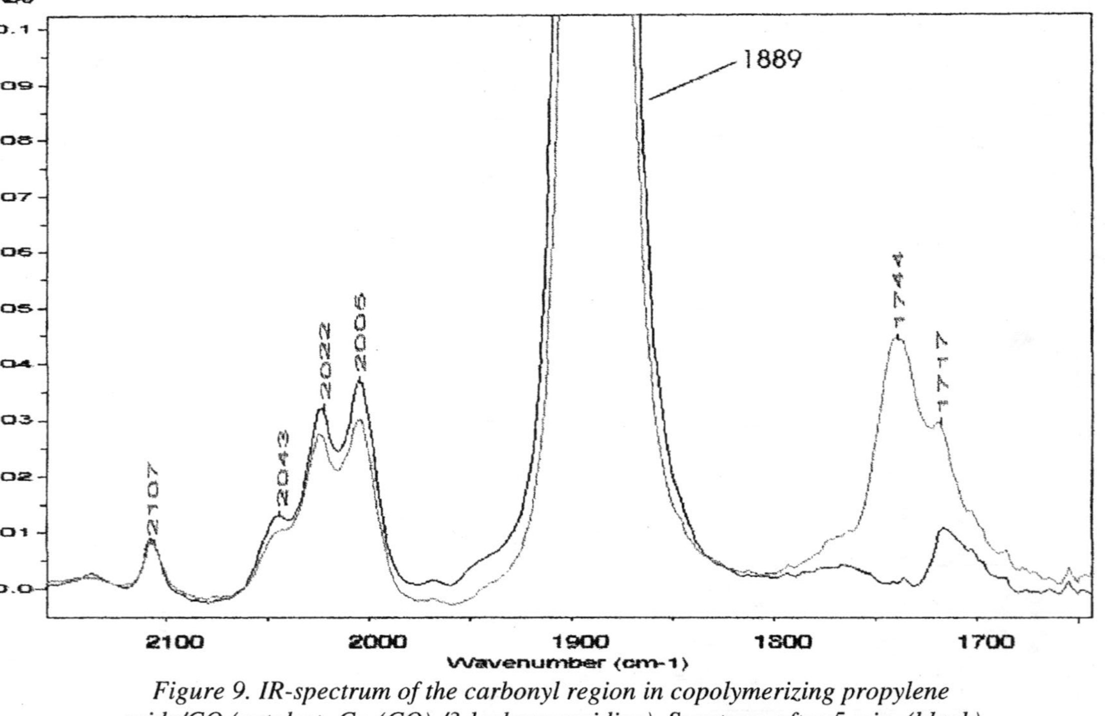

Figure 9. IR-spectrum of the carbonyl region in copolymerizing propylene oxide/CO (catalyst: $Co_2(CO)_8$/3-hydroxypyridine). Spectrum after 5 min. (black) and after 20 min. (red). $Co(CO)_4^-$ (1889 cm^{-1}), $Co(CO)_4$(acyl) 2107, 2043, 2022, 2005, 1717 cm^{-1}, polyhydroxybutyrate 1744 cm^{-1}.

(Reproduced from reference 15. Copyright 2002 American Chemical Society.)

CO pressure did not result in the formation of $HCo(CO)_4$, rather a fast first epoxide ring opening was observed followed by reversible CO insertion, yielding the acyl complex 3 (Figure 10; 2107, 2043, 2022, 2003, 1710 cm^{-1}; black to red). In analogy to Heck, no further reaction, especially no polymer formation is observed at this point (*25*).

The addition of a simple base, like pyridine (3-hydroxypyridine is not necessary) can now be used to induce a spontaneous formation of polyhydroxybutyrate, indicated by a growing absorption around 1744 cm^{-1} (Figure 10, red to blue). A presumption on the role of pyridine in the formation of PHB is based on a cleavage of the Co-acyl bond by C_5H_5N to form a kind of ion pair acylpyridinium - tetracarbonyl cobaltate (3a→3b, Figure 11) (*26*). Pyridine is known to catalyze reactions of activated acyl compounds through N-acyliminium (pyridinium) ions (*27*). Pyridine is probably a suitable mediator for these reactions because it is an efficient nucleophile, smoothly reacting with acyl compounds that bear a good leaving group (i.e. $Co(CO)_4^-$) (*28*). In addition, an acylated pyridinium does not easily loose a proton. It is to be expected, that the pyridinium acyl is highly reactive towards—epoxide monomers. Thus it is proposed, that the acylpyridinium - tetracarbonyl cobaltate species reacts with PO, liberating pyridine and forming a β-carboxyalkyl cobaltate complex. A cationic ring opened PO building block of constitution acyl-OCH(Me)-CH$_2$+ is a likely intermediate that can combine with $Co(CO)_4^-$ to give the corresponding cobalt alkyl complex. From this point, the polyester formation can proceed through alternating CO and pyridine induced epoxide insertions (*29*).

Conclusions

It has been shown that the synthesis of polyhydroxyalkanoates can be achieved through reaction of epoxides and CO under the action of a tetracarbonyl cobaltate/pyridine catalyst mixture. No lactones are intermediately formed. Most likely an alternating copolymerization takes place with the following reaction steps: (i) Lewis acid induced epoxide ring opening, (ii) capture of the resulting carbocationic alkyl by tetracarbonyl cobaltate to give a β-alkylcobalt carbonyl, (iii) CO insertion yielding an acyl-cobalt complex, and (iv) pyridine induced formation and stabilization of an acylium moiety that again promotes epoxide ring opening.

Acknowledgement

We are grateful to the Bundesministerium für Bildung und Forschung (BMBF, Grant 03C310) for generous financial support.

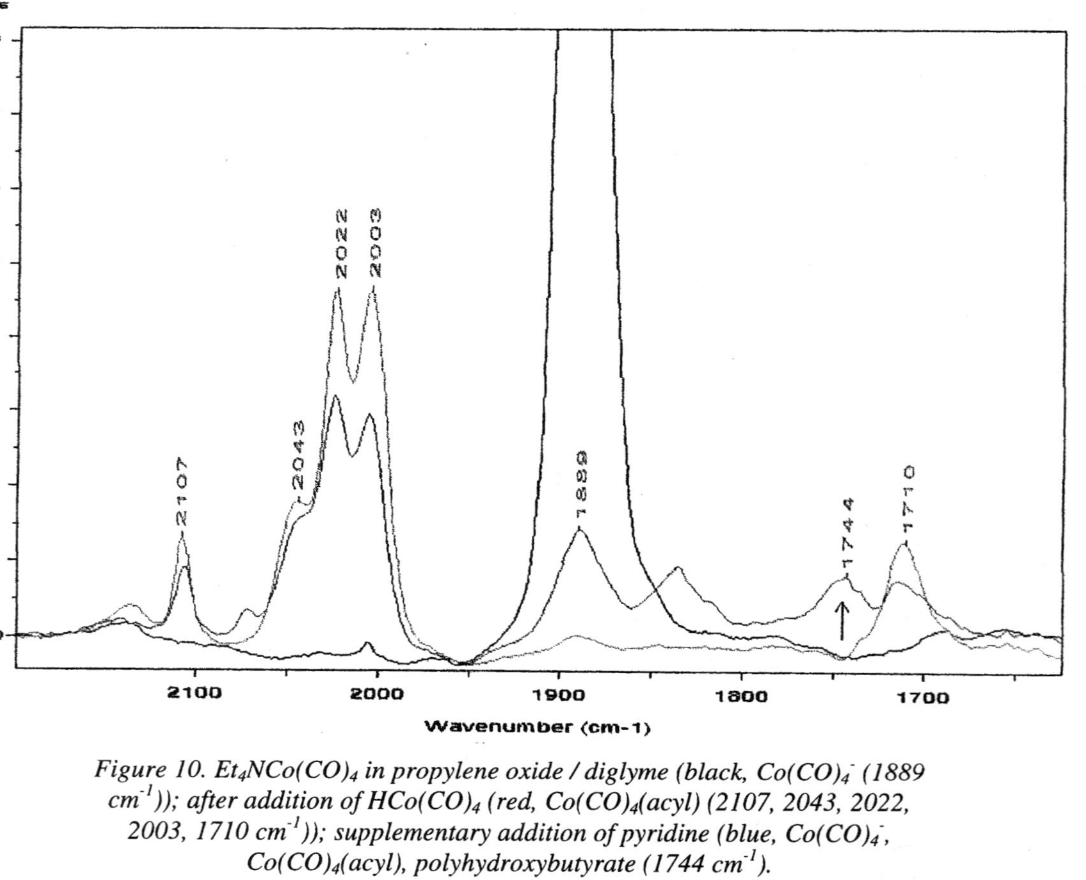

Figure 10. Et₄NCo(CO)₄ in propylene oxide / diglyme (black, Co(CO)₄⁻ (1889 cm⁻¹)); after addition of HCo(CO)₄ (red, Co(CO)₄(acyl) (2107, 2043, 2022, 2003, 1710 cm⁻¹)); supplementary addition of pyridine (blue, Co(CO)₄⁻, Co(CO)₄(acyl), polyhydroxybutyrate (1744 cm⁻¹).

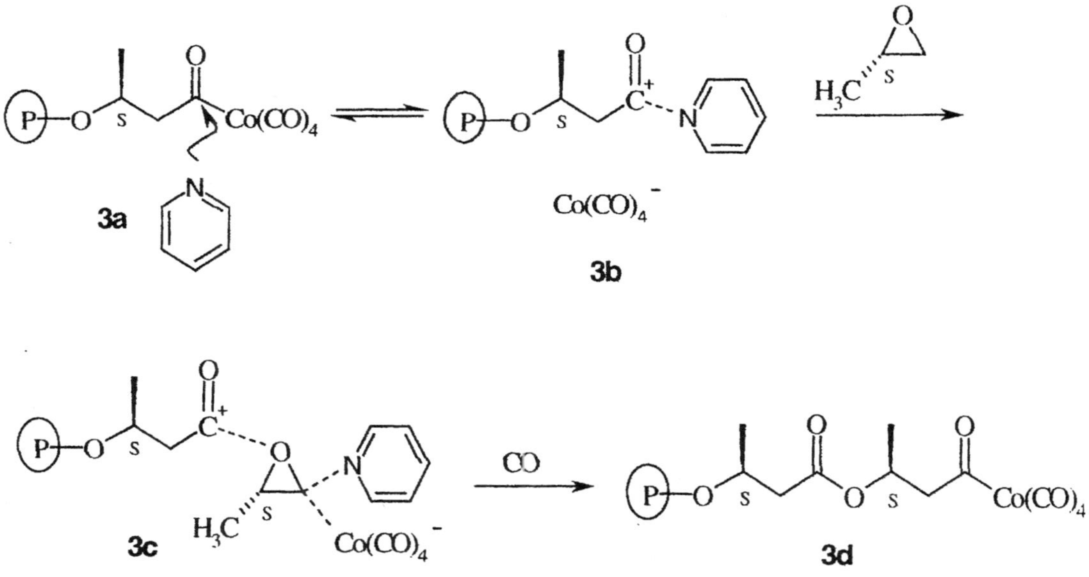

Figure 11. Proposed mechanism for the epoxide insertion into a cobaltcarbonyl-acyl bond by pyridine mediation.

128

1. (a) Lafferty, R.M.; Korsatko, B.; Korsatko, W. In *Biotechnology*; Rehm, H.J., Ed.; Verlag Chemie: Weinheim, Germany, 1988, Vol 6b. (b) Trau, M.; Truss, R.W. *Eur. Pat. Appl.* EP 293 172 A2, 1988. (c) Reeve, M.S.; McCarthy, S.; Gross, R.A. *Polym. Prepr.* **1990**, 31(1), 437.

2. (a) Doi, Y.; Microbial Polyesters, VCH, New York **1990**. (b) *Novel Biodegradable Mirobial Polyesters*; NATO ASI Series; Daves, E.A., Ed.; Kluwer Academic Publishers: Dordrecht, The Netherlands, 1990. (c) Marchessault, R.H. *Polym. Prepr.* **1988**, 29, 584-585. (d) Swift, G. *Acc. Chem. Res.* **1993**, 26, 105-110. (e) Anderson, A. J.; Dawes, E. A. *Macrobiol. Rev.* **1990**, 54, 450.

3. (a) Lemoigne, M.C.R. *Acad. Sci.* **1925**, 180, 1539. (b) Lemoigne, M.C.R. *Bull. Soc. Chim. Biol.* **1926**, 8, 1291.

4. (a) Dawes, E.A.; Senior, P. J. *Adv. Microbiol. Phys.* **1973**, 10, 138. (b) Müller, H. M.; Seebach, D. *Angew. Chem. Int. Ed. Engl.* **1993**, 32, 477. (c) Pool R. *Science* **1989**, 245, 1187.

5. (a) Kunioka, M.; Doi, Y. *Macromolecules* **1990**, 23, 1923. (b) Grassie, N.; Murray, E.J.; Holmes, P.A. *Polym. Degrad. Stab.* **1984**, 6, 47/95/127.

6. (a) Holmes, P.A.; *Phys. Technol.* **1985**, 16, 32. (b) Schmitt, E.E.; US Patent 3 297 033. (c) Lowe C.E. U.S. Patent 1 668 162. (d) Holmes, P.A.; Wright, L.F.; Collins, S.H. Eur. Pat. Appl. 0 052 459, 1992; Eur. Pat. Appl. 0 069 497, 1982. (e) *BIOPOL, a degradable polymer* in: Gummi, Fasern, Kunststoffe **1991**, 44, 451-452.

7. (a) Bluhm, T.L.; Hamer, G.K.; Marchessault, R.H.; Fyfe, C.A.; Veregin, R.P. *Macromolecules* **1986**, 19, 2871. (b) Bloembergen, S.; Holden, D.A.; Bluhm, T.L.; Marchessault, R.H. *Macromolecules* **1989**, 22, 1663.

8. (a) Jedlinski, Z.; Kowalczuk, M. *Macromolecules* **1989**, 22, 3242. (b) Le Borgne, A.; Pluta, Ch.; Spassky, N. *Macromol. Rap. Commun.* **1994**, 15, 955-960. (c) Hori, Y.; Suzuki, M.; Yamaguchi, A.; Nishishita, T. *Macromolecules* **1993**, 26, 5533. (d) Kricheldorf, H. R.; Eggerstedt S. *Macromolecules* **1997**, 30, 5693-5697.

9. (a) Getzler, Y. D. Y. L.; Mahadevan V.; Lobkovsky, E. B.; Coates G. W. *J. Am. Chem. Soc.* **2002**, 124, 1174. (b) Lee, T. L.; Thomas, P. J.; Alper, H. *J. Org. Chem.* **2001**, 66, 5424-5426. (c) Molnar, F.; Luinstra, G.A.; Allmendinger, M.; Rieger, B. *Chem. Eur. J.*, in press.

10. (a) Arcana, M.; Giani-Beaune, O.; Schue, F.; Amass, W.; Amass, A. *Polym. Int.* **2000**, 49, 1348-1355; (b) Zhang, Y.; Gross, R. A.; Lenz, R. W. *Macromolecules* **1990**, 23, 3206 ; (c) Le Borgne, A.; Spassky, N. *Polymer* **1989**, 30, 2312 ; (d) Abe, H.; Matsubara, I.; Doi, Y.; Hori, Y.; Yamaguchi, A. *Macromolecules* **1994**, 27, 6018.

11. (a) Ittel, S.D.; Johnson, L.K.; Brookhart, M. *Chem. Rev.* **2000**, 100, 1169-1203. (b) Brintzinger, H.H.; Fischer, D.; Mülhaupt, R.; Rieger, B.; Waymouth, R. M. *Angew. Chem., Int. Ed. Engl.* **1995**, 34, 1143.

12. (a) Drent, E.; Budzelaar, H.M. *Chem. Rev.* **1996**, 96, 663. (b) Sommazzi, A.; Garbassi, F. *Prog. Polym. Sci.* **1997**, 22, 1547-1605. (c) Abu-Surrah, A.S.; Rieger, B. *Topics in Catalysis* **1999**, 7, 165.

13. (a) Cheng, M.; Moore, D. R.; Reczek, J. J.; Chamberlain, B. M.; Lobkovsky, E. B.; Coates, G. W. *J. Am. Chem. Soc.*; **2001**; 123(36); 8738-8749. (b) Rokicki, A.; Kuran, W. J. *Macromol. Sci. Rev., Macromol. Chem.* **1981**, C21, 135. (c) Inoue, I *Chemtech* **1976**, 6, 588. (d) Mang, S.; Holmes, A.B. *Macromolecules* **2000**, 33, 303-308.

14. (a) Hinterding, K.; Jacobsen, E.N. *J. Org. Chem.* **1999**, 64, 2164-2165. (b) Penney, J. M. *Diss. Abstr. Int., B* **1999**, 60, 2688. (c) Kamiya, Y.; Kawato, K.; Ohta, H. *Chem. Lett.* **1980**, 1549-1552. (d) Takegami, Y.; Watanabe, Y.; Mitsudo, T.; Masada, H. *Bull. Chem. Soc. Jpn.* **1968**, 41, 158. (e) Heck, R.F.; *J. Am. Chem. Soc.* **1963**, 85, 1460. (f) Eisenmann, J.L.; Yamartino, I.L.; Howard, J.F. *J. Org. Chem.* **1961**, 26, 2102.

15. Allmendinger, M.; Eberhardt, R.; Luinstra, G.A; Rieger, B. *J. Am. Chem. Soc.* **2002**, 125, 5646-7.

16. We repeated this reaction with propylene oxide and analyzed it by online ATR-IR. The produced compounds are a mixture of β-butyrolactone and poly-hydroxybutyrate in equal quantities and some further unidentified products in lower amounts. Furukawa, J.; Iseda, Y.; Fujii, H. *Macromol. Chem.* **1965**, 89, 263.

17. Drent, E.; Kragtwijk, E. Eur. Pat. Appl. EP 577, 206, 1994.

18. ATR IR spectrometer: ReactIR^{TM}, SiComp^{TM} probe from Mettler Toledo for insitu-IR measurements under high pressure conditions in a 250ml Büchi reactor.

19. The decreased concentration of the lactone after addition of propylene oxide (Figure 6, B) results from dilution of the mixture.

20. (a) Iida, M.; Hayase, S.; Araki, T. *Macromolecules* **1978**, 11, 490. (b) Gross, R. A.; Zhang, Y.; Konrad, G.; Lenz, R. W. *Macromolecules* **1988**, 21, 2657. (c) Bloembergen, S.; Marchessault, R.H. *Macromolecules* **1989**, 22, 1656. (d) Le Borgne, A.; Spassky, N. *Polymer* **1989**, 30, 2312-2319.

21. Kemmit, R. D. W.; In *Comprehensive Organometallic Chemistry* Wilkinson, G., Ed.; Pergamon Press: New York 1982; Vol. 5, 34.2.4.

22. $HCo(CO)_4$ is a relatively strong acid in aquaeous media. However, there is only limited information on pk_s-values in organic solution. We cannot detect larger quantities of $HCo(CO)_4$ from the reaction of 3HO-Py and $Co_2(CO)_8$. Therefore we assume that the role of H^+ is to polarize the epoxide monomer and to act as polymer "end group".

130

23. Allmendinger, M.; Eberhardt, R.; Luinstra, G.A; Rieger, B. *PMSE Preprints* **2002**, 86, 332.

24. (a) Heck, R.F.; *J. Am. Chem. Soc.* **1961**, 83, 4023. (b) Heck, R.F. *Adv. Organomet. Chem.* **1966**, 4, 243. (c) Heck, R.F., in: *Organic Syntheses via Metal Carbonyls* Wender, I.; Pion, P., Ed.; Wiley Interscience: New York 1968; Vol. 1, 373. (d) Heck, R.F., in: *Organotransition Metal Chemistry* Academic Press: New York 1974; 201. (e) Kreisz, J.; Ungváry, F.; Markó, L. *J. Organomet. Chem.* **1991**, 417, 89. (f) Kovács, I.; Ungváry, F. *Coord. Chem. Rev.* **1997**, 161, 1-32.

25. Alternatively it is possible to start here with any other acyl cobalt-carbonyl complex, synthesized insitu or before.

26. Complementary, a step growth mechanism might be considered. Thereby a Lewis acid, like a Co^{2+}(py)-species could induce the epoxide ring opening (similar to H^+) together with a $Co(CO)_4^-$-anion. After carbonylation, the resulting polyester building block would be transfered to a growing polyester chain, leading to an elongation by one segment. However, DFT calculations show that there is a low energy pathway according to Figure 11. Detailed theoretic and experimental results will be published elsewhere.

27. (a) Satchell, D. P. N.; Satchell, R. S In *Supplement B: The Chemistry of Acid Derivatives*; Patai, S., Ed.; Wiley: New York, 1992, Volume 2, Chapter 13. (b) Hiemstra, H.; Speckamp, W. N. In *Comprehensive Organic Synthesis: Additions to N-Acyliminium Ions*; Trost, B. M.; Fleming, I., Ed.; Pergamon Press: Oxford 1991, 1047-1082.

28. (a) Fersht, A.R.; Jencks, W.P. *J. Am. Chem. Soc.* **1970**, 92, 5432-5442. (b) Page, M.; Williams, A. *Organic and Bioorganic Mechanisms*; Longman: Harlow, 1997; pp 145-152. (c) Imyanitov, N.S. *Kinet. Catal.* **1999**, 40, 71-75.

29. There is also a chance that pyridine might assist in the electrophilic attack of the acyl carbon atom on the epoxide (Figure 11). This hypothesis is supported by a recent publication on a similar effect in Zn-catalyzed epoxide/CO_2-coupling reactions: Kim, H.S.; Kim, J.J.; Kang, S.O. *Angew. Chem.* **2000**, 112, 22, 4262.

Chapter 10

Copolymerization of Ethylene and Acrylates by Nickel Catalysts

Lynda Johnson[1], Lin Wang[1], Steve McLain[1], Alison Bennett[1], Kerwin Dobbs[1], Elisabeth Hauptman[1], Alex Ionkin[1], Steven Ittel[1], Keith Kunitsky[1], William Marshall[1], Elizabeth McCord[1], Catherine Radzewich[1], Amy Rinehart[1], K. Jeff Sweetman[1], Ying Wang[1], Zuohong Yin[1], and Maurice Brookhart[2]

**[1]DuPont Central Research and Development, Experimental Station, Wilmington, DE 19880–0328
[2]Department of Chemistry, University of North Carolina, Chapel Hill, NC 27599–3290**

Ethylene/acrylate copolymerizations can be carried out with α-diimine nickel complexes and with other nickel catalysts with productivities significantly higher than those previously reported with palladium. The key factor in enabling this reaction with the nickel catalysts is the use of higher temperatures and, to some extent, higher pressures. The effect of varying the ligand substituents and the reaction parameters is discussed along with details regarding scouting procedures, copolymer characterization, and the reaction mechanism. Details regarding the synthesis and reactivity of new phosphine-based nickel catalysts are also given.

Introduction

Virtually all catalysts for the commercial polymerization of *nonpolar* olefins are based on early transition metals such as titanium, zirconium and chromium. Due to their high oxophilicity, these early transition metal catalysts are typically poisoned by *polar* olefins such as acrylates, and therefore, copolymers of polar olefins with ethylene are still produced commercially using free radical initiators under high-pressure and high-temperature conditions in expensive reactors.

In 1996, it was reported that palladium α-diimine complexes (Versipol®catalysts) catalyze the copolymerization of ethylene and acrylates to give random copolymers (*1,2*). These Pd-catalyzed copolymerizations are typically carried out at low ethylene pressures and low temperatures (e.g., 1 atm ethylene and ambient temperature) with higher pressures and temperatures resulting in decreased acrylate incorporation and catalyst decomposition, respectively. In contrast, polar olefin copolymerizations with nickel catalysts have been limited primarily to polar α-olefins in which a number of methylenes separate the polar and vinyl groups (*3,4*). Reaction of Ni catalysts with olefins containing polar functionality directly bound to the vinyl group has typically resulted in catalyst deactivation (*5*).

We report here that ethylene/acrylate copolymerizations can be carried out with α-diimine Ni catalysts and with other Ni catalysts with good productivities (*6-14*). The key enabling factor is the use of higher temperatures and, to some extent, higher pressures.

Copolymerization Reactions with α-Diimine Ni Complexes

Due to ease of synthesis and stability, Ni allyl initiators were typically used for copolymerization screening reactions, which were carried out in small 30 mL reactors. Dependent upon the catalyst and the reaction conditions, acrylate homopolymerization is sometimes a significant side reaction, which can be limited by the addition of salts such as NaBAF (BAF = B[3,5-$C_6H_3(CF_3)_2$]$_4$) and LiB(C_6F_5)$_4$. Screening with higher acrylates such as ethylene glycol phenyl ether acrylate is useful for rapidly and quantitatively analyzing by ^{1}H NMR spectroscopy for copolymer formation in the presence of poly(acrylate) formation and residual monomer.

A number of α-diimine Ni complexes were tested for copolymerization activity, with those based upon the acenaphthene backbone exhibiting some of the highest productivities. Yields were highest when the copolymerization reactions were carried out at elevated temperatures (~80 °C and above) and pressures (~500 psi and above). Although copolymerizations could be carried out in the absence of any cocatalyst, the addition of a Lewis acid such as $B(C_6F_5)_3$ often improved the productivity.

The effect of varying temperature, pressure and steric bulk on productivity, percent acrylate incorporation, branching and molecular weight is illustrated in Table I with some representative ethylene (E) and methyl acrylate (MA) copolymerization results obtained with α-diimine Ni catalysts **1** and **2** (eq 1). The copolymerizations were carried out with quite low amounts of catalyst (0.0019 mmol) and long reaction times (90 h). Identical copolymerizations with shorter reaction times gave lower yields, indicating that these copolymerizations are long-lived.

Table I. Ethylene/Methyl Acrylate Copolymerizations[a]

Cmpd	Press psi	Temp °C	Yield g	Incorp. mol%	M_w	Total Me^b
1	500	120	0.89	1.33	7,201	72.5
2	500	120	0.84	0.93	9,799	94.9
1	1000	120	3.34	0.82	13,853	50.4
2	1000	120	5.42	0.43	30,997	68.5
1	1000	100	3.87	0.62	24,270	30.9
2	1000	100	4.05	0.51	50,877	48.0

[a]0.0019 mmol Cmpd, 211 equiv $B(C_6F_5)_3$, 105 equiv NaBAF, 90 h, 0.5 mL MA, 9.5 mL p-xylene. [b]Total Me/1000 CH_2.

^{1}H and ^{13}C NMR spectroscopy and dual UV/RI detection GPC supported the formation of random copolymers of ethylene and acrylate. Dependent on the reaction conditions and the nature and bulk of the α-diimine substituents, the copolymers ranged from linear to highly branched with predominantly in-chain acrylate incorporation. In contrast, the ester groups of ethylene/acrylate copolymers made by the analogous Pd α-diimine catalysts occur largely at the ends of branches.

Experimental Design: E/MA Copolymerization with Diimine Ni Catalyst 1

An experimental design in a larger reactor was undertaken to further elucidate the effects of the variables temperature, ethylene pressure, acrylate concentration, and $B(C_6F_5)_3$ cocatalyst concentration on E/MA copolymerizations utilizing catalyst **1**.

Polymerization Procedure

All polymerizations for the experimental design were carried out in a 450 mL jacketed autoclave with steam/water temperature control as follows: In a drybox, 1.77 g $Na[B(C_6H_3(CF_3)_2)_4]$, $B(C_6F_5)_3$ (100, 200, or 300 equiv per Ni), methyl acrylate (5, 12.5, or 20 volume % - sparged with nitrogen and passed through a column of activated neutral alumina in the drybox immediately before use), and all but 20 mL of the toluene were combined. The total volume of the reaction mixture was 100 mL and the amount of methyl acrylate and toluene were calculated on this basis. This solution was transferred to a metal addition cylinder in the drybox, and the solution was then charged to the nitrogen-purged autoclave. In a drybox, the Ni catalyst **1** (0.0286 g) was dissolved in the remaining 20 mL of toluene and mixed for 30 minutes on a lab vibramixer. This solution was transferred via cannula under positive nitrogen pressure to a small metal addition cylinder attached to the autoclave. While stirring, the autoclave was charged with ethylene to 50 psig and vented three times. The autoclave and its contents were heated to the reaction temperature (100, 120, or 140 °C). After achieving the desired operating temperature, the autoclave was pressurized with ethylene to 50 to 100 psig below the desired operating pressure. The catalyst addition cylinder was charged with ethylene to the desired operating pressure, and the catalyst solution was then pressure injected to begin the polymerization. Ethylene was fed to maintain a constant pressure. After 6 hours, the reaction was cooled to room temperature. All volatiles were removed from the reaction solution using a rotary evaporator. The nonvolatile material was washed with three 50-100 mL portions of methanol with decanting of the methanol into a fritted glass filter after each wash. The methanol insoluble solids were then transferred to the filter with methanol and washed on the filter with additional methanol. The solids were dried for 18 hours at 60 °C in a vacuum oven. Molecular weights of the isolated polymer samples were determined by GPC. ^{13}C NMR spectroscopy of the dried polymers was used to calculate the weight % of homo-poly(methyl acrylate) which was subtracted from the total yield to determine copolymer yield, the mole % MA content of the ethylene/methyl acrylate copolymer, and the amount of hydrocarbon branching in the copolymer. Polymerizations for kinetic measurements were run at 120 °C, 600 psig ethylene, 5 volume % methyl acrylate, and 200 equivalents of $B(C_6F_6)_3$. Except for reaction time, the polymerization procedure was identical to other experiments.

Data Analysis

The experimental design was set up and analyzed using ECHIP$^©$ software. To minimize the number of experiments we chose a three level interactive response surface design that uses only first order terms and pairwise interactions, and is specially constructed to detect lack of fit against quadratic models. The design required 21 total experiments including 5 replicate pairs.

Experimental Design

The Pareto effects graph for copolymer yield (Figure 1) shows that temperature, methyl acrylate volume % and their cross term are the three most important variables. In the Pareto effects graph, a "*" indicates a positive parameter and a "o" indicates a negative parameter. Although temperature and methyl acrylate volume % are the dominant variables, the model is complex and all of the variables and their cross terms have a statistically significant effect on yield. The importance of the cross terms reveals significant interactions between variables. The interaction model obtained from the experimental design gave an excellent fit to the copolymer yield data as evidenced by the linearity of the plot in Figure 2.

Despite the interactions between variables, it proved possible to rank the relative importance of each individual variable on copolymer yield, methyl acrylate incorporation, degree of branching from chain-walking, and molecular weight (Table II). The numbers within a given column reflect the rank order of the importance of each variable on that polymer property (1>2, etc.), and the +/- signs indicate whether the variable has a positive or negative effect on that polymer property. A major conclusion can be drawn from comparing the signs in the columns for methyl acrylate incorporation and molecular weight. For each variable, the signs in these two columns are opposite. This suggests that it will not be possible to achieve both high molecular weight and high methyl acrylate incorporation with this catalyst. Scanning the columns of Table II, it is clear that the variables methyl acrylate volume % and temperature have large effects on polymer properties, whereas the amount of $B(C_6F_5)_3$ has only minor effects.

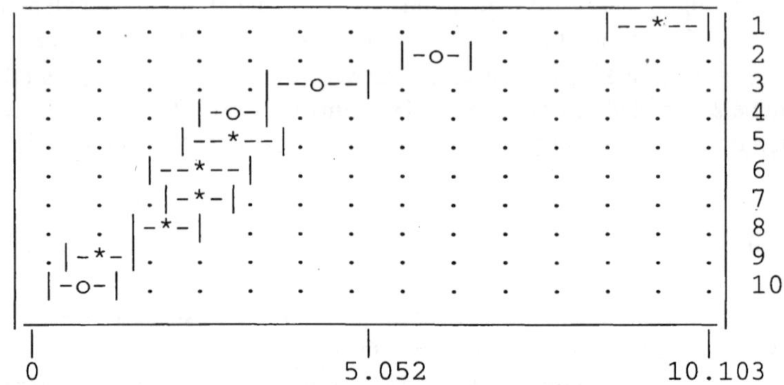

Figure 1. Pareto effects graph for copolymer yield.

Key: 1=temperature, 2=methyl acrylate, 3=(temperature)(methyl acrylate), 4=(methyl acrylate)($B(C_6F_5)_3$), 5=(temperature)($B(C_6F_5)_3$), 6= $B(C_6F_5)_3$, 7=ethylene, 8=(ethylene)($B(C_6F_5)_3$), 9=(temperature)(ethylene), 10=(methyl acrylate)(ethylene)

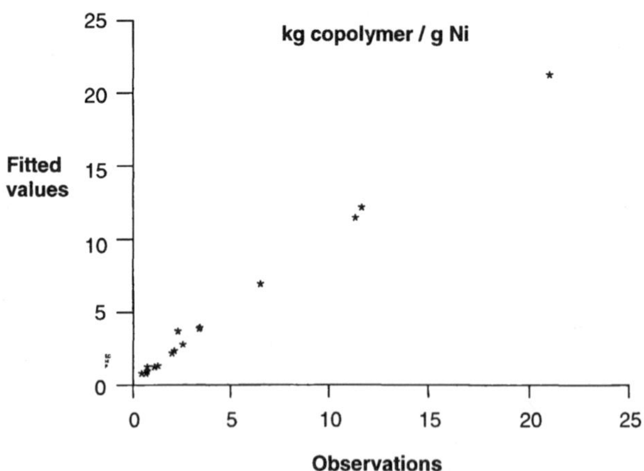

Figure 2. Copolymer yield – fitted vs. observed values.

Table II. Summary of Major Effects

Variable	Yield	Incorp (mol%)	Total Me[a]	Mol. Wt. (M_w)
MA (vol%)	2 (-)	*1 (+)*	No effect	*1 (-)*
Temp. (° C)	1 (+)	3 (+)	1 (+)	2 (-)
$B(C_6F_5)_3$	3 (+)	$(+)^b$	No effect	$(-)^b$
Ethylene (psig)	4 (+)	2 (-)	No effect	3 (+)

[a]Total Me/1000 CH_2. [b]Borderline statistical significance.

Copolymerization Kinetics

The reaction rate and profile were obtained by running copolymerizations for 3, 6, and 10 hours, and plotting the copolymer yield vs. time (Figure 3). The data was fit to a kinetic model that assumes that the rate of catalyst deactivation is first order in Ni. The catalyst half-life obtained from this fit is 3.5 hours. It should be noted that at this temperature with this catalyst, an ethylene homopolymerization would have a half-life of only a few minutes. Although methyl acrylate slows the rate of polymerization, it clearly improves the thermal stability of the catalyst. This is thought to be due to the chelating effect of the newly enchained methyl acrylate (*vide infra*).

Mechanistic Studies

Low-temperature NMR studies (eq's 2 and 3) investigating the reaction of α-diimine Ni complex **3** with MA showed that MA insertion occurs at -40 °C. The acrylate complex was not observed. A 4-membered chelate complex **4** formed initially and slowly rearranged at room temperature to a 6-membered chelate complex **5** within the course of one day. This is indicative of a strong Ni-carbonyl interaction as, in comparison, the rearrangement of the analogous Pd 4-membered chelate complex occurs at -80 °C (*1*).

Based upon these studies, elevated temperatures and pressures aid the dissociation of the carbonyl from the electrophilic Ni metal center. The Lewis acid cocatalyst may serve to bind to the dissociated carbonyl and hold it off of the metal center long enough for ethylene to bind and insert, thus continuing the catalytic cycle. Although the strong chelate interaction slows the rate of

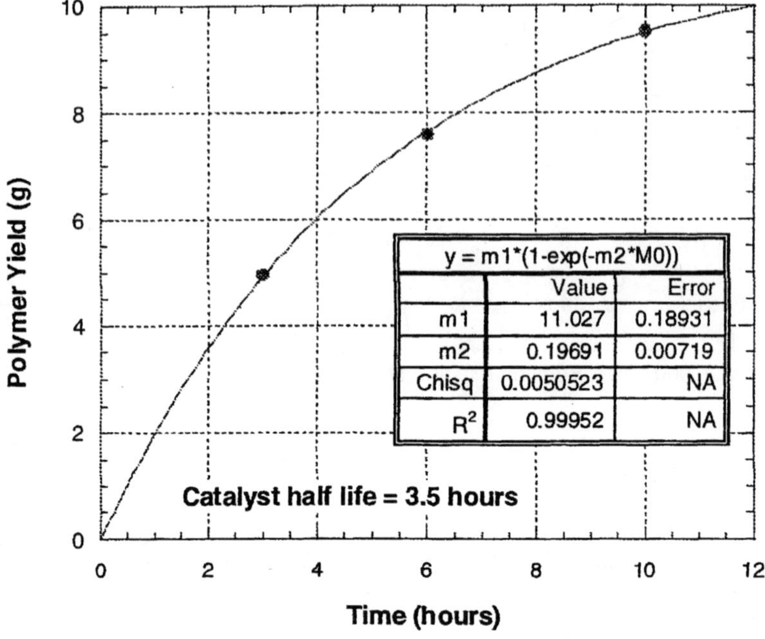

Figure 3. Copolymer yield vs. time.

ethylene/acrylate copolymerization, it has the favorable effect of slowing catalyst decomposition, resulting in long half-lives as compared to ethylene homopolymerizations with these Ni catalysts.

Copolymerizations with Phosphine-Based Ni Catalysts

The general utility of higher temperature and pressure conditions for Ni-catalyzed ethylene/acrylate copolymerizations has been demonstrated with a number of different Ni catalysts in addition to the α-diimine systems, including a large number of new phosphine-based neutral and Zwitter-ionic nickel catalysts. As illustrated in eqs 4 and 5, the new phosphine-based ligands were readily synthesized by using strong nucleophiles such as $(t\text{-}Bu)_2PCH_2Li$ and $(t\text{-}Bu)_2PLi$ to attack electrophiles (**E**) such as ketones, imines, epoxides, isocyanates, nitriles, carbodiimides, carbon dioxide and other unsaturated organic molecules by addition across the C=N and C=O bonds or opening of the epoxide rings. The ligand precursors were subsequently allowed to react with half an equivalent of allyl nickel chloride dimer to lead to the formation of the phosphine-based nickel catalysts.

Some examples of the nickel catalysts made through this chemistry are listed in Figure 4. Structures of complexes **6-12**, **19** and **20** were confirmed by X-ray crystal structure analyses. Complexes **8-11** each existed as a dimer bridged through LiCl in the solid state. The Zwitter-ionic complexes **19** and **20** were made by reacting compounds **17** and **18** with $B(C_6F_5)_3$ and $B[3,5\text{-}C_6H_3\text{-}(CF_3)_2]_3$, respectively. Detailed synthetic procedures may be found in the patent application (*10*).

Polymerization and Polymer Microstructure

Upon activation with Lewis acids such as $B(C_6F_5)_3$, these phosphine-based nickel catalysts were found to be active for ethylene, ethylene/α-olefin, and ethylene/polar olefin (co)polymerizations, including the copolymerization of acrylates (Table III). The Zwitter-ionic catalysts **19** and **20** were active for olefin polymerization in the absence of additional Lewis acid cocatalysts.

Polyethylenes made by these catalysts in general had a low degree of alkyl branching, indicating that olefin insertion was highly favored over chain-walking for these catalysts. Nonetheless, NMR analyses of the branching distributions were consistent with a chain-walking mechanism similar to that observed for the α-diimine-based catalysts (*1-3*). Polymers made by catalyst **6** often exhibited no alkyl branches and high molecular weights.

The ethylene/acrylate copolymers made by these new phosphine-based nickel catalysts also had a low degree of alkyl branching. The ester groups of these polymers were generally directly connected to the polymer backbone (IC) or located in end groups of the polymer main chain (EG), rather than being located at the end of the side chains.

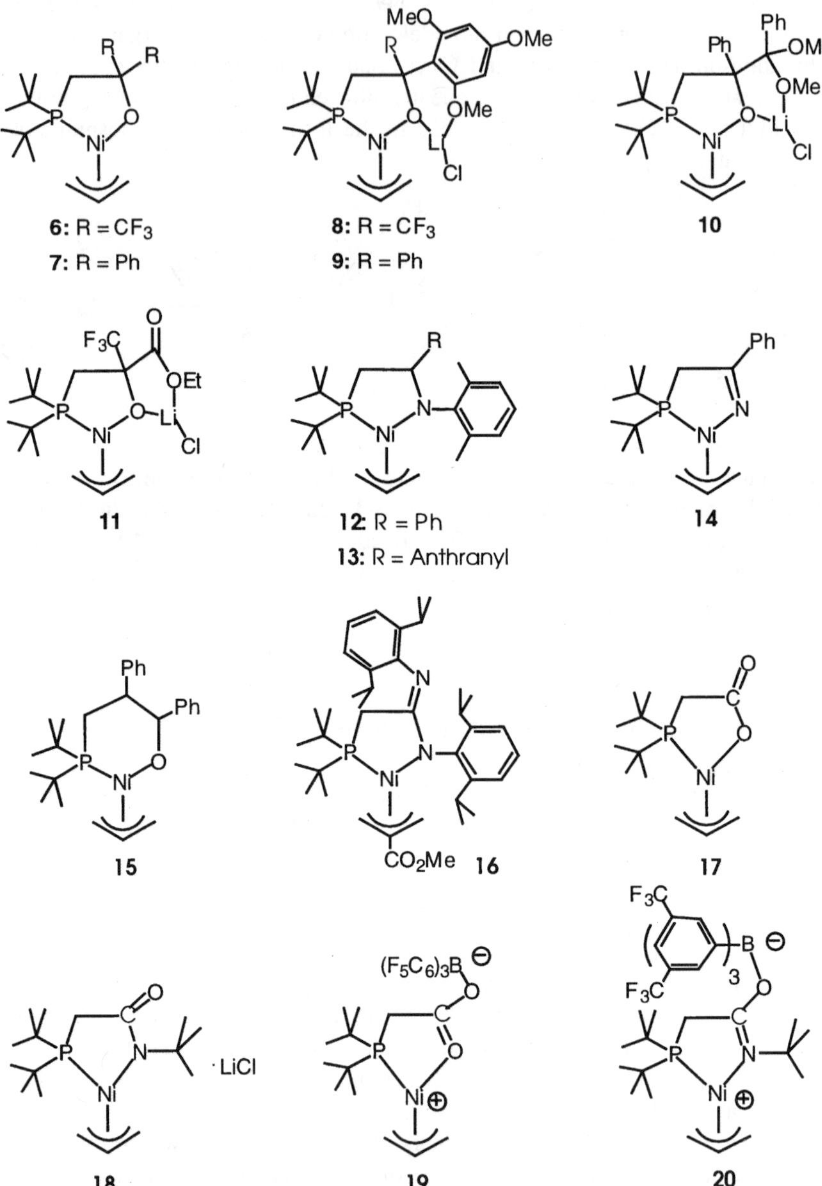

Figure 4. Phosphine-based nickel catalysts.

Table III. Ethylene/Hexyl Acrylate (HA) Copolymerizations[a]

Cmpd	Yield (g)	Total Me[b]	Incorp. (mol%)	m.p. (°C)	M_w	PDI[c]
7	0.461	9	0.41 0.20IC 0.21EG	125	22,311	10.8
8	3.998	10	0.54 0.25IC 0.29EG	124	3,567	2.2
9	1.253	3	0.53 0.25IC 0.28EG	122	11,123	10.8
10	1.075	19	0.38 0.19IC 0.19EG	112	1,405	1.8

[a]0.02 mmole Cmpd, 40 equiv $B(C_6F_5)_3$, 20 equiv $LiB(C_6F_5)_4$, 8 mL TCB, 2 mL HA, 120 °C, 1000 psi ethylene, 18 h. [b]Total Me/1000 CH_2. [c]PDI: Polydispersity index.

Catalysts Incorporating Lewis Acids

As shown in Table III, the LiCl-coordinated catalysts **8-10** were more productive than catalyst **7** in ethylene/hexyl acrylate copolymerizations. A similar trend was observed for ethylene homopolymerization. Likely, the incorporation of LiCl in **8-10** renders the metal center more electrophilic, improving the activity in olefin polymerization. To further enhance the electrophilicity of the catalysts **8-10**, $LiB(C_6F_5)_4$ was added to the reaction mixture (see Table III). In a separate experiment, NaBAF was added to catalyst **8** in ether, resulting in the abstraction of the chloride ion and the formation of ether coordinated cationic complex **21** (eq 6). The structure of **21** was confirmed by X-ray single crystal analysis.

Upon activation with Lewis acids such as $B(C_6F_5)_3$, Catalysts **17** and **18** were very active for ethylene/acrylate copolymerization. The active form of the catalyst is likely Zwitterionic (*e.g.,* catalysts **19** and **20**). For example, catalyst **18** produced 28 kg of copolymer per gram of Ni with 0.41 mole percent MA incorporation (0.02 mmol **18**, 1000 psi E, 100 °C ,18 h, 9.5 mL TCB, 0.5 mL MA, 20 equiv $B(C_6F_5)_3$, 20 equiv NaBAF, yield = 13.21 g, incorp: 0.21 mol% IC and 0.20 mol% EG, 9 Me/1000 CH_2, M_w=5,008, PDI=3.2).

References

1. Johnson, L. K.; Mecking, S.; Brookhart, M. *J. Am. Chem. Soc.* **1996**, *118*, 267.
2. Mecking, S.; Johnson, L. K.; Wang, L.; Brookhart, M. *J. Am. Chem. Soc.* **1998**, *120*, 888.
3. Ittel, S. D.; Johnson, L. K.; Brookhart, M. *Chem. Rev.* **2000**, *100*, 1169.
4. Boffa, L. S.; Novak, B. M. *Chem. Rev.* **2000**, *100*, 1479.
5. Suzuki, Y.; Hayashi, T. JP 11292918 (Mitsui), 1998. This application reports ethylene/acrylate copolymerizations catalyzed by α-diimine Ni catalysts at low pressures and temperatures. The reported characterization of the copolymer structure is very limited, however, and productivities and percent acrylate incorporations are low.
6. The text, figures and tables in this chapter are adapted from references 7 - 9. Additional experimental details may be found in references 10 - 14.
7. Johnson, L.; Bennett, A.; Dobbs, K.; Hauptman, E.; Ionkin, A.; Ittel, S.; McCord, E.; McLain, S.; Radzewich, C.; Yin, Z.; Wang, L.; Wang Y.; Brookhart, M. *PMSE Preprints* **2002**, *86*, 319.
8. McLain, S. J.; Sweetman, K. J.; Johnson, L. K.; McCord, E. *PMSE Preprints* **2002**, *86*, 320.
9. Wang, L.; Hauptman, E.; Johnson, L. K.; Marshall, W. J.; McCord, E. F.; Wang, Y.; Ittel, S. D.; Radzewich, C. E.; Kunitsky, K.; Ionkin, A. S. *PMSE Preprints* **2002**, *86*, 322.
10. Wang, L.; Hauptman, E.; Johnson, L. K.; McCord, E. F.; Wang, Y.; Ittel, S. WO 0192342 (DuPont), 2000.
11. Johnson, L. K.; Wang, L.; McCord, E. F. WO 0192354 (DuPont), 2000.
12. Johnson, L. K.; Bennett, A. M. A.; Dobbs, K. D.; Ionkin, A. S.; Ittel, S. D.; Wang, Y.; Radzewich, C. E.; Wang, L. WO 0192347 (DuPont), 2000.
13. Wang, L.; Johnson, L. K.; Ionkin, A. S. WO 0192348 (DuPont), 2000.
14. Brookhart, M. S.; Kunitsky, K.; Malinoski, J. M.; Wang, L.; Wang, Y.; Liu, W.; Johnson, L. K.; Kreutzer, K. A.; Ittel, S. D. WO 0259165 (DuPont), 2000.

Chapter 11

Trends in Alkene Insertion in Late- and Early-Transition Metal Compounds: Relevance to Transition Metal-Catalyzed Polymerization of Polar Vinyl Monomers

Myeongsoon Kang[1], Ayusman Sen[1,*], Lev Zakharov[2], and Arnold L. Rheingold[2]

[1]Department of Chemistry, The Pennsylvania State University, University Park, PA 16802
[2]Department of Chemistry, University of Delaware, Newark, DE 19716

Variable-temperature ^1H NMR studies of the reaction of cationic (α-diimine)Pd-methyl complex with alkenes are presented. The studies reveal that vinyl bromide coordinates to the Pd(II)-methyl complex, followed by migratory insertion and β-bromo elimination, to generate free propene. Propene further reacts to give β-agostic Pd(II)-*tert*-butyl species. From the reactions with vinyl halide, stable chloro-bridged dicationic palladium complex was isolated and characterized. For a series of alkenes ($CH_2=CHX$), the rate for migratory insertion decreases as follows : X = CO_2Me > Br > H > Me.

144

Introduction

Metal-catalyzed insertion copolymerization of polar vinyl monomers with nonpolar alkenes remains an area of great interest in synthetic polymer chemistry, because the addition of functionalities to a polymer which is otherwise non-polar can greatly enhance the range of attainable properties (*1*). For vinyl monomers with pendant coordinating functionalities, such as acrylates, the principal problem has been catalyst poisoning through functional group coordination (*1-3*). Interestingly, vinyl halides, which do not posses any strongly coordinating functionality, are also not polymerized by any known transition metal-catalyzed systems. Recently, Jordan (*4*) and Wolczanski (*5*) have reported the reaction of vinyl halides with *rac*-(EBI)ZrMe$_2$ and (^{t}Bu$_3$SiO)$_3$TaH$_2$, respectively. It was demonstrated that, following 1,2-insertion of the alkene, β-halide elimination occurs to generate a metal-halide bond. Since the halophilicity of the transition metal ions tends to decrease on going from left to right in the periodic table, we undertook an examination of the reactivity of vinyl halides towards a late transition metal complex (palladium). Herein we report mechanistic investigations of the reactions of cationic (α-diimine)Pd-alkyl complexes with alkenes using variable-temperature ^{1}H NMR spectroscopy (*6*).

Results

Generation of Active Species. The starting point of our investigation was the Brookhart-type cationic Pd(II)-methyl species, **2**, generated in our case by the addition of two equivalents of AlCl$_3$ to the corresponding neutral Pd(II)-methylchloride, **1**, (Schemes 1 and 2). ^{1}H NMR studies revealed that the active species **2** forms through the intermediate **A** or **B** (Scheme 1) (*7*). The reaction of **1** and 1 equiv. of AlCl$_3$ resulted in the generation of **A/B** and trace amounts of **2** upon stirring for 2 h at room temperature. The addition of a second equiv. of AlCl$_3$ to the reaction mixture resulted in the formation of the active species **2**. The ^{1}H NMR spectrum of **2** is similar to that of the Brookhart compound [(ArN=C(Me)C(Me)=NAr)Pd(Me)(NCMe)]BAr'F (Ar = 2,6-di-*i*-Pr-C$_6$H$_3$), implying similar structures for both. However, the Brookhart compound was unreactive towards vinyl bromide, presumably because of the failure of the olefin to displace the coordinated MeCN. Utilization of the Brookhart's system for the investigation of the reaction pathway using vinyl bromide was unsuccessful.

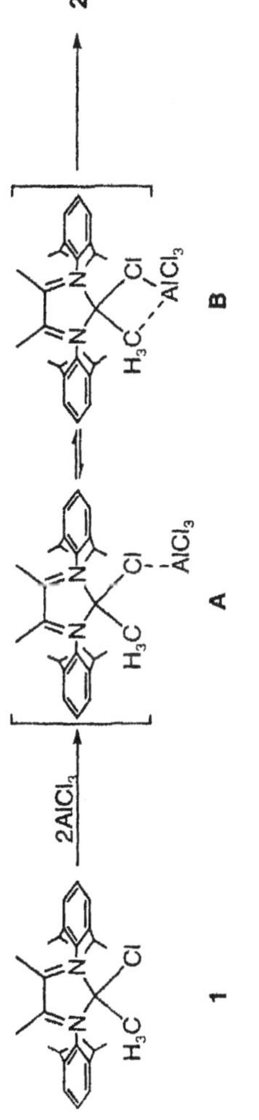

Scheme 1. Generation of active species (7)

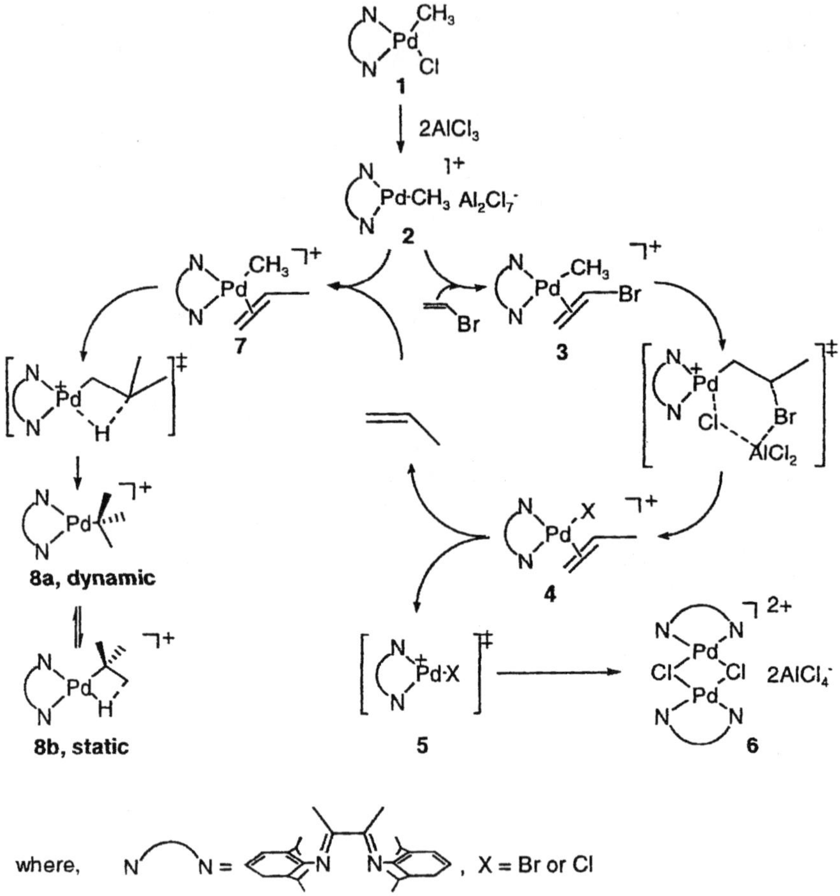

146

$2AlCl_3$

1

2

3

4

5

6

7

8a, dynamic

8b, static

where, N͡N = , X = Br or Cl

Scheme 2. Proposed Reaction Pathway (6,7)

¹H NMR Study of the Reaction of Cationic (α-diimine)Pd-methyl Complex with Vinyl bromide. Several equivalents of vinyl bromide were added to a CD_2Cl_2 solution of **2** at -90°C and the reaction mixture was monitored by ¹H-NMR spectroscopy as it was gradually warmed up (Scheme 2). The coordination of vinyl bromide to the metal center in **2** was observed even at -86°C, resulting in the formation of **3**. Warming the reaction mixture to -74°C resulted in the formation of the propene coordinated species, **4**, suggesting 1,2-alkene insertion followed by β-bromo elimination. Propene is gradually lost from **4** and is trapped by unreacted **2** to form **7**. Formation of complex **7** was confirmed by addition of free propene to the system, which caused the ratio of **7** to **4** to increase. The cationic Pd(II)-halide species arising from **4** by propene loss converts to the chloro-bridged dimer **6**. The structure of **6** as a dicationic complex with two aluminum tetrachloride counteranions was established by an X-ray crystal structure determination (Figure 1).

Once formed, **7** undergoes 1,2-insertion of propene to form the known β-agostic Pd(II)-*tert*-butyl compound **8** (*8*). The complexes **7** and **8** were also independently formed by the addition of propene to a CD_2Cl_2 solution of **2**. The initially formed **7** was found to convert to **8** when the solution was warmed to -36°C. At ambient temperature, the three methyl groups of the *tert*-butyl complex exchange rapidly on the NMR time scale and appear as a singlet at –0.28 ppm. Upon lowering the temperature to -86°C, a static ¹H NMR spectrum is observed and the agostic proton appears as a singlet at –8 ppm.

Kinetics of the Reaction of Cationic (α-diimine)Pd-methyl Complex with Alkenes. The migratory insertion rates of bound vinyl bromide and propene in **3** and **7**, respectively, were directly measured by monitoring the disappearance of the corresponding Pd-CH_3 resonance. For propene our value was in close agreement with that reported by Brookhart (*10*). For vinyl bromide, an Arrhenius plot was constructed from rate measurements done between -74 and -37°C. Our values together with those of Brookhart (*9,10*) are reported in Table 1.

A Hammett plot of the relative insertion rates of substituted alkenes versus σ_p (*19*) yielded a straight line with a *positive* ρ (+3.41) (Fig. 2). Of note is that the line encompasses values obtained by both Brookhart (*9,10*) and us.

Discussion

Formation of Dimer and Deactivation Pathway. The identity of the halide ligand in **4** (Scheme 2) has not been established but the formation of **6** opens up the possibility of an aluminum-assisted β-bromo abstraction pathway shown in Scheme 2. The dimer formation is observed only when vinyl halides,

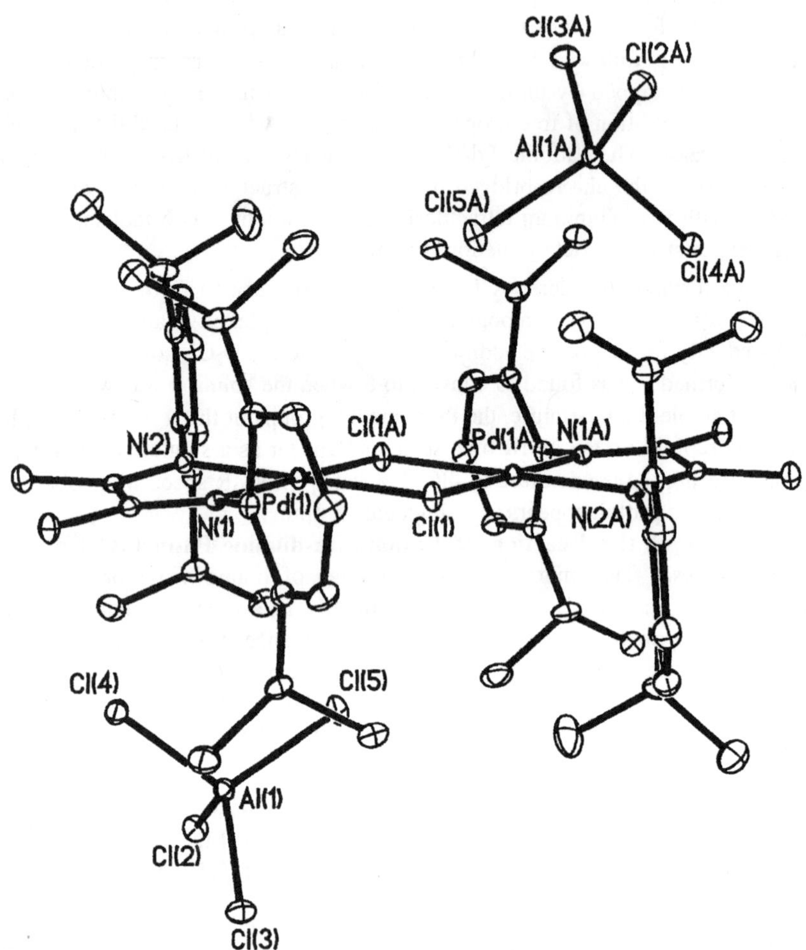

Figure 1. ORTEP view of **6**. *Hydrogen atoms are not shown for clarity*

Table 1. Kinetic Data for Insertion of Alkenes into Pd(II)-Me bond[a]

	$k\ (x\ 10^3\ s^{-1})$	$\Delta H^{\ddagger}\ (kcal/mol)$	$\Delta S^{\ddagger}\ (cal/K{\bullet}mol)$
Propene	0.54		
Ethene[10]	1.9	14.2 ± 0.1	-11.2 ± 0.8
Vinyl Bromide	22.0	11.9 ± 0.1	-16.8 ± 0.1
Methyl Acrylate[9]	55.0	12.1 ± 1.4	-14.1 ± 7.0

[a] Measured or extrapolated to 236.5K

(*Reproduced from* reference 6. *Copyright 2002 American Chemical Society.*)

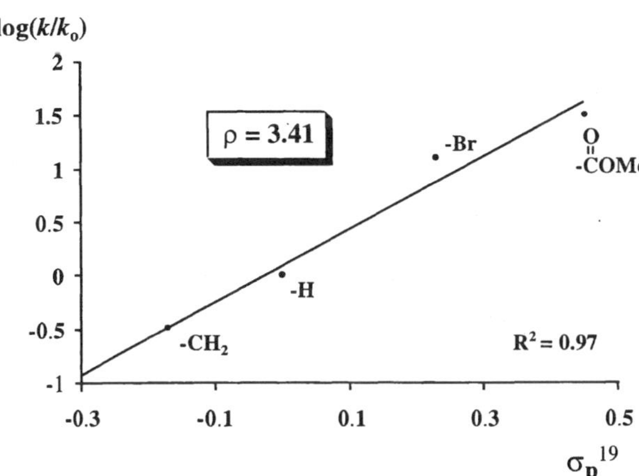

Figure 2. *Hammett plot for migratory insertion of alkenes into palladium(II)-methyl bond.*

Source: Reproduced from reference 6. Copyright 2002 American Chemical Society.

150

such as vinyl bromide or vinylidene chloride, are utilized as monomers. Surprisingly, β-bromopropyl-Pd(II) complex was not observed. It is likely that the β-bromoalkyl-metal intermediate rapidly reacts with excess aluminum chloride, followed by Cl/Br exchange, to form the dichloro-bridged palladium-diimine complex **6** which crystallizes out of solution due to its low solubility.

Prolonging the reaction time of **1** with 2 equiv. of $AlCl_3$ in CD_2Cl_2 also generates **4**, which implies another possible pathway for the formation of the dichloro-bridged dimer, **6**.

Relative Insertion Rates of Alkenes. We have observed a linear *positive* Hammett correlation between the rate of insertion and the increasing electron-withdrawing effect of the substituent on the alkene. Theoretical calculations also suggest a lower insertion barrier for acrylates (but not vinyl halides) compared to ethene (*11,12*). Interestingly, we did not find evidence for the theoretically predicted 2,1-insertion of vinyl halide into the palladium-methyl bond (*12*). Our results can be contrasted with a *negative* value of ρ obtained by Bercaw for alkene insertion in $Cp*_2NbH(alkene)$ (*13*). Additionally, The second-order rate constants obtained by Wolczanski for alkene insertion into Ta(V)-H bond follow the trend H ≈ OR >> halide (F, Cl, Br) (*5*). The decrease in the rate of insertion with increasing electron withdrawing effect of the substituent on the alkene in case of early transition metal compounds has been attributed to the development of positive charge on the carbon bearing the substituent either during alkene coordination or the subsequent insertion step (*5,13,14*). In Wolczanski's case, it is not been possible to separate the effect of the substituent on binding versus insertion, and the trend for the actual insertion step remains an open question. We ascribe the increase in insertion rate for the palladium-methyl complex to a ground-state effect. An alkene with an electron-withdrawing substituent coordinates less strongly to the electrophilic metal (i.e. σ-donation is more important than π-back-donation) (*15*). Thus, a weaker metal-alkene bond has to be broken for the insertion to proceed (i.e., the destabilization of the alkene complex leads to a lower insertion barrier). Another surprising observation is that the observed correlation extends to propene. The slower insertion and polymerization rates of propene, when compared to ethene, is usually attributed to the steric bulk of the methyl group of the former (*10,16-18*). Our results suggest that, at least for the late transition metal compounds, it is the donating ability of the methyl group, rather than its size, which results in slower rate for propene insertion and, hence, polymerization. This argument applies even for the sterically encumbered Brookhart-type system. Likewise, for acrylates, the precoordination of the ester group appears to have little effect on its migratory insertion rate.

Conclusions

What is the implication of our work with respect to the metal-catalyzed polymerization of polar vinyl monomers? First, for the late metal compounds, the polar vinyl monomers can clearly outcompete ethene and simple 1-alkenes with respect to insertion. However, the ground-state destabilization of the alkene complex that favors the migratory insertion of the polar vinyl monomers is a two-edges sword because it biases the alkene coordination towards ethene and 1-alkenes. Indeed, we have observed the near quantitative displacement of vinyl bromide by propene to form **7** from **3**. Thus, the extent of incorporation of the polar vinyl monomer in the polymer will depend on the opposing trends in alkene coordination and migratory insertion. The above discussion does not take into account the problem of functional group coordination for acrylates or β-halide abstraction for vinyl halides. With respect to the latter, we are currently exploring approaches to suppress this "termination" step, e.g. decreasing the electrophilicity of the metal center.

Experimental Section

All work involving air and moisture sensitive compounds was carried out under an inert atmosphere of argon or nitrogen by using standard Schlenk or glove-box techniques. Variable-temperature ^{1}H NMR experiments were performed on a Brüker DPX 300 NMR spectrometer, using CD_2Cl_2 as solvent. NMR probe temperatures were measured using an Omega type T thermocouple immersed in anhydrous methanol in a 5 mm NMR tube.

All solvents were deoxygenated, dried via passage over a column of activated alumina, degassed by repeated freeze-pump-thaw cycles, and stored over 4 Å molecular sieves under nitrogen. The α-diimine ligand, 2,6-$C_6H_4(i$-Pr$)_2$-N=C(Me)-C(Me)=N-2,6-$C_6H_4(i$-Pr$)_2$, and complex, (2,6-$C_6H_4(i$-Pr$)_2$-N=C(Me)-C(Me)=N-2,6-$C_6H_4(i$-Pr$)_2$)Pd(Me)(Cl) were prepared as previously reported (*20,21*). AlCl$_3$ (Strem) was stored in an nitrogen-filled drybox and used as received. Vinyl bromide (98%) was purchased from Aldrich and used without further purification.

Rates for migratory insertion of alkenes into the Pd(II)-methyl bond were determined by adding 20 equiv of alkenes to the pre-generated complex **2** solution in CD_2Cl_2 and monitoring the loss of the Pd(II)-CH$_3$ resonance over time. The natural logarithm of the methyl integral was plotted versus time (first-order treatment) to obtain kinetic plots. Activation parameters ($\mathit{\Delta G}^{\ddagger}$, $\mathit{\Delta H}^{\ddagger}$, and $\mathit{\Delta S}^{\ddagger}$) were calculated from measured rate constants and temperatures using the Eyring equation.

References

1. Boffa, L. S.; Novak, B. M. *Chem. Rev.* **2000**, *100*, 1479-1493.

2. Brintzinger, H. H.; Fischer, D.; Mulhaupt, R.; Rieger, B.; Waymouth, R. M. *Angew. Chem. Int. Ed. Engl.* **1995**, *34*, 1143-1170.

3. Coates, G. W.; Waymouth, R. M. In *Comprehensive Organometallic Chemistry II*, Abel, E. W.; Stone, F. G. A.; Wilkinson, G., Eds.; Elsevier: New York, **1995**; Vol. 12, p. 1193.

4. Stockland, R. A. Jr.; Jordan, R. F. *J. Am. Chem. Soc.* **2000**, *122*, 6315-6316.

5. Strazisar, S. A.; Wolczanski, P. T. *J. Am. Chem. Soc.* **2001**, *123*, 4728-4740.

6. Kang, M.; Sen, A.; Zakharov, L.; Rheingold, A. L. *J. Am. Chem. Soc.,* **2002**, *124*, 12080-12081.

7. Selected ^{1}H NMR data (CD_2Cl_2, 300MHz, δ in ppm) (a) ^{1}H NMR of **A/B** (25°C): 2.88 and 2.74 (septet, 2H each, Ar-C*H*MeMe'), 2.06 and 2.02 (s, 3H each, N=C(C*H$_3$*)-C(C*H$_3$'*)=N), 1.21, 1.13, 1.08, and 1.00 (d, 6H each, Ar-CHC*H$_3$*C*H$_3$'*), 0.39 (s, 3H, Pd-C*H$_3$*). (b) ^{1}H NMR of **2** (25°C): 7.42~7.34 (m, 6H, Ar-*H*), 3.04 (septets, 4H, Ar-C*H*MeMe'), 2.22 and 2.17 (s, 3H each, N=C(C*H$_3$*)-C(C*H$_3$'*)=N), 0.76 (s, 3H, Pd-C*H$_3$*). (c) ^{1}H NMR of **3** (-70°C): 5.58 (m, 1H, coordinated CH$_2$=C*H*Br), 5.0 (m, 2H, coordinated C*H$_2$*=CHBr), 2.44 and 2.33 (s, 3H each, N=C(C*H$_3$*)-C(C*H$_3$'*)=N), 0.65 (s, 3H, Pd-C*H$_3$*). (d) ^{1}H NMR of **4** (-50°C): 5.47 (m, 1H, coordinated CH$_2$=C*H*CH$_3$), 4.21 (m, 2H, coordinated C*H$_2$*=CHCH$_3$), 2.00 (m, coordinated CH$_2$=CHC*H$_3$*). (e) ^{1}H NMR of **7** (-40°C): 4.31 (m, 2H, coordinated C*H$_2$*=CHCH$_3$) CH$_2$=C*H*CH$_3$ obscured, CH$_2$=CHC*H$_3$* obscured, 0.32 (s, 3H, Pd-C*H$_3$*) (f) ^{1}H NMR of **8a**, dynamic (-20°C): −0.27 (br s, 9H, Pd(C(C*H$_3$*)$_3$)), ^{1}H NMR of **8b**, static (-70°C): 0.00 (br s, 6H, Pd(C(CH$_2$-μ-H)-(C*H$_3$*)$_2$)), −8.00 (br s, 1H, Pd(C(CH$_2$-*μ-H*)-(CH$_3$)$_2$)). (g) ^{1}H NMR of **6** (25°C): 2.33 (s, 12H, N=C(C*H$_3$*)-C(C*H$_3$'*)=N), 1.35 and 1.23 (d, 24H each, J = 6.59 and 6.86 Hz).

8. Shultz, L. H.; Tempel, D. J.; Brookhart, M. *J. Am. Chem. Soc.* **2001**, *123*, 11539-11555.

9. Mecking, S.; Johnson, L. K.; Wang, L.; Brookhart, M. *J. Am. Chem. Soc.* **1998**, *120*, 888-899.

10. Tempel, D. J.; Johnson, L. K.; Huff, R. L.; White, P. S.; Brookhart, M. *J. Am. Chem. Soc.* **2000**, *122*, 6686-6700.

11. Michalak, A.; Ziegler, T. *J. Am. Chem. Soc.* **2001**, *123*, 12266-12278.

12. Phillipp, D. M.; Muller, R. P.; Goddard, W. A.; Storer, J.; McAdon, M.; Mullins, M. *J. Am. Chem. Soc.* **2002**, *124*, 10198-10210.

13. Doherty, N. M.; Bercaw, J. E. *J. Am. Chem. Soc.* **1985**, *107*, 2670-2682.
14. Burger, B. J.; Santarsiero, B. D.; Trimmer, M. S.; Bercaw, J. E. *J. Am. Chem. Soc.* **1988**, *110*, 3134-3146.
15. von Schenck, H.; Strömberg, S.; Zetterberg, K.; Ludwig, M.; Åkermark, B.; Svensson, M *Organometallics*, **2001**, *20*, 2813-2819.
16. Michalak, A.; Ziegler, T. *Organometallics*, **1999**, *18*, 3998-4004.
17. Michalak, A.; Ziegler, T. *Organometallics*, **2000**, *19*, 1850-1858.
18. Michalak, A.; Ziegler, T. *Organometallics*, **2001**, *20*, 1521-1531.
19. Hansch, C.; Leo, A.; Taft, R. W. *Chem. Rev.* **1991**, *91*, 165-195.
20. Rülke, R. E.; Ernsting, J. M.; Spek, A. L.; Elsevier, C. J.; van Leeuwen, P. W. N. M.; Vrieze, K. *Inorg. Chem.* **1993**, *32*, 5769-5778
21. van Asselt, R.; Elsevier, C. J.; Smeets, W. J. J.; Spek, A. L.; Benedix, R. *Recl. Trav. Chim. Pays-Bas* **1994**, *113*,88-98.

Chapter 12

Theoretical Studies on the Polymerization and Copolymerization Processes Catalyzed by the Late-Transition Metal Complexes

Artur Michalak[1,2] and Tom Ziegler[1]

[1]Department of Chemistry, University of Calgary, 2500 University Drive, NW, Calgary, Alberta T2N 1N4, Canada
[2]Department of Theoretical Chemistry, Jagiellonian University, Ul. R. Ingardena 3, 30–060 Cracow, Poland

In the present account we address two aspects of the polymerization processes catalyzed by the late-transition-metal complexes: (i) influence of the catalyst structure and the reaction conditions (temperature and olefin pressure) on the polyolefin microstructure; (ii) co-polymerization of α-olefins with oxygen-containing monomers. Gradient corrected DFT has been used to determine the energetics and the activation barriers of the elementary reactions in these processes. A stochastic model has been developed and used to simulate the polymer growth under different T, p conditions.

Introduction

Although the traditional Zigler-Natta and the early-transition-metal catalysts presently dominate industrial polymerization processes, there has been an increasing interest in the development of catalytic systems based on the late-transition-metal complexes (*1, 2*). Similarly to the early-metal-based systems, the late-metal-complexes can yield high molecular-weight polymers. Moreover, they can lead to polyolefins with different microstructures, depending on the catalysts and the reaction conditions (temperature, olefin pressure) (*2-6*). Unlike the heterogeneous catalysts and the early-metal-complexes, the late-metal-based systems exhibit substantial tolerance toward polar groups, and as such they may be useful for the co-polymerization of α-olefins with polar monomers (*1, 2, 7-9*).

In the present account we address two aspects of the polymerization processes catalyzed by the late-transition-metal complexes:

(i) influence of the catalyst structure and the reaction conditions (temperature and olefin pressure) on the polyolefin microstructure (topology);

(ii) co-polymerization of α-olefins with the oxygen-containing monomers.

Recently, hyperbranched polymers were obtained (*2-6*) in the polymerizations of simple monomers such as ethylene and linear α-olefins, catalyzed by Ni- and Pd-diimine catalysts (*2,10,11*). Branches in these polyolefins form as a result of fast chain isomerization reactions. The topology of the polymers is strongly affected by the olefin pressure: under low pressure highly-branched structures are obtained, whereas high pressure gives rise to structures with linear side-chains. Interestingly, in the polymerization catalyzed by Pd-diimine complexes the average number of branches is pressure independent, while for the Ni-based system it is strongly affected by the pressure. Branching in olefin polymerization catalyzed by diimine complexes can be controlled also to some extent by the polymerization temperature and the substituents on the catalyst. Recently, another late transition metal catalyst has been shown to exhibit similar branching features: in the polymerization processes catalyzed by a neutral Ni-anilinotropone complex polymer branching can be controlled by a variation in temperature and pressure (*12*).

Quantum chemistry has establish itself as a valuable tool in the studies of polymerization processes (*13,14*). However, direct quantum chemical studies on the relationship between the catalyst structure and the topology of the resulting polymer, as well as on the influence of the reaction conditions are not practical without the aid of statistical methods. We have to this end proposed a combined approach in which quantum chemical methods are used to provide information on the microscopic energetics of elementary reactions in the catalytic cycle, that is required for a 'mesoscopic' stochastic simulations of polymer growth (*15*). A stochastic approach makes it possible to discuss the effects of temperature and olefin pressure.

The major goal of this study was to understand the factors contoling polyolefin branching and the relationship between the catalyst structure, temperature, pressure and the polyolefin topology. The DFT calculations were carried out for the elementary reactions in the polymerization of ethylene and propylene catalyzed by Pd-based diimine catalysts (*16,17*) and the ethylene polymerization catalyzed by the Ni-anilinotropone catalyst (*18*). The polymer growth in these processes was modeled by a stochastic approach (*15,18,19*). Further, the model simulations were performed, by systematically changing insertion barriers, to model the influence of catalyst, beyond the diimine systems (*19*).

The design of an efficient catalyst for the copolymerization of olefins with polar group containing monomers is one of the challenges for the polymerization chemistry (*2,7-9*) The studies in (ii) have been focused on understanding the factors responsible for the differences between the Pd- (active catalyst for ethylene-acrylate copolymerization) and Ni-diimine (inactive) catalyst (*2,7-9*). The full mechanistic DFT studies, involving static calculations and molecular dynamic (MD) simulations have been performed for both Ni- and Pd-based systems (*20-22*). The results of this study reveal the factors that inhibit the polar copolymerization in the case of the Ni-diimine catalyst.

Modeling the polyolefin branching

Scheme 1 presents the mechanism of the α-olefin polymerization catalyzed by late-metal catalysts. In these process, the resting state of the catalyst in an olefin π-complex, **A**, from which the polymer chain may grow via 1,2- (**RA**) or 2,1- inserion (**RB**). Both insertion paths introduce one methyl branch. The "chain straightening" isomerization reaction (**RC**) is responsible for a removal of a branch; this isomerization reaction may follow the 2,1- insertion. However, the isomerisation reactions may also elongate branches, when they proceed in the opposite direction (**RD, RE**, etc.). Another isomerisation reaction which may follow the 1,2-insertion introduces an additional methyl branch, and also may proceed further (**RF, RG**, etc.). Thus, in the polymerization cycle, many different alkyl species are present (**B-F**), in which the metal atom forms a bond with primary, secondary, or tertiary carbon atoms; each of them can capture a new monomer and give rise to a subsequent insertion. Thus, in order to understand the influence of the catalyst and reaction conditions on the polymer microstructure, one must consider all these elementary reactions.

Energetics of elementary reactions / relative stability of isomers

In two recent papers (*16*,17) we reported the results of computational studies on the elementary reactions in the ethylene and propylene polymerization catalyzed by Pd-based diimine catalysts with different substituents (Scheme 2). Comparison of the results for the model catalysts (**1**) and the real systems (**2-7**) makes it possible to separate the electronic and steric effects. Energetics of the elementary reactions in the catalytic cycle obtained from the calculations for **7** is summarized and compared with experimental data (*23-25*) in Figure 1. Similar studies were performed (*18*) for the ethylene polymerization with the Ni-anilinotropone complex.

Relative stability of isomeric alkyl complexes and the olefin π-complexes

The calculations were performed for the isomeric propyl and butyl complexes. Two general trends can be observed (*16-18*) for both, diimine and anilinotropone complexes: (i) the more branched the alkyl is, the more stable the

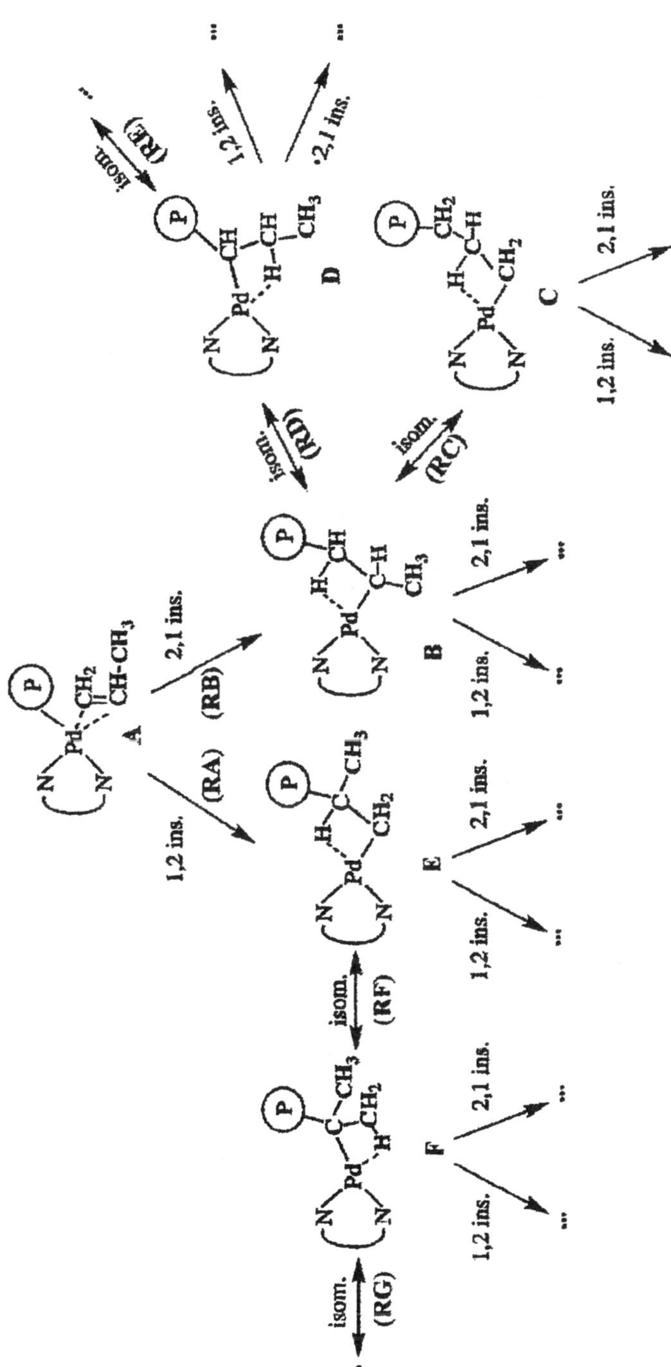

Scheme 1. Chain growth and isomerization reactions in the α-olefin polymerization catalyzed by Pd-diimine catalyst

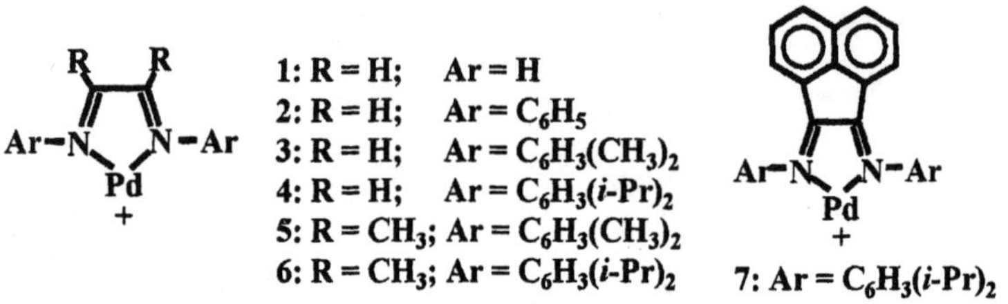

1: R = H; Ar = H
2: R = H; Ar = C_6H_5
3: R = H; Ar = $C_6H_3(CH_3)_2$
4: R = H; Ar = $C_6H_3(i\text{-}Pr)_2$
5: R = CH_3; Ar = $C_6H_3(CH_3)_2$
6: R = CH_3; Ar = $C_6H_3(i\text{-}Pr)_2$
7: Ar = $C_6H_3(i\text{-}Pr)_2$

Scheme 2. Model and 'real' Pd(II)-based diimine catalysts

Primary alkyl
Secondary alkyl
TS (A→A')
TS (B→C)
TS (B'→C')
7.2
(5.8*)
1.0 (0.5)
ΔE1
ΔE2
A
1.3
(1.6*)
A'
18.5
(18.9)
17.4
(17.4)
B
B'
2.1
(2.0)

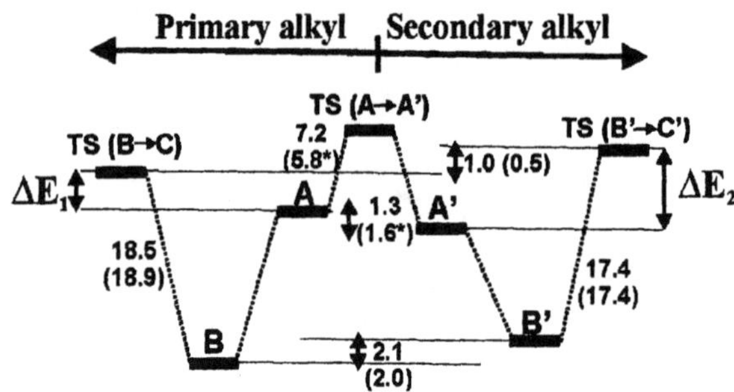

*Figure 1.Energetics [in kcal /mol] of the chain growth and chain isomerization reactions in the ethylene polymerization catalyzed by the Pd-diimine complex 7 (see Scheme 2). The experimental and theoretical (in parantheses) values are presented. The calculated energies for the chain isomerization reactions (marked with *) were obtained from the model catalyst 1.*

corresponding β-agostic complex is; (iii) the steric bulk on the catalyst has little effect on the relative stability of the alkyl complexes. There are two opposing factors determining the stability of alkyl complexes: (i) the stability of free alkyl radicals, increasing for more branched systems; (ii) the energy of binding of the radicals, decreasing for more branched alkyls (*17*). As a result, the energy order of isomeric alkyl complexes resembles the energy order of the alkyl radicals, with smaller energy differences between them than between the free radicals.

For the model catalysts **1** the olefin π-complexes with branched alkyls are more stable than with the linear ones (*16*). This electronic preference is strongly affected by the steric bulk on the catalyst due to an interaction between the alkyl group and the catalyst substituents (*17*). Thus, for the most bulky real catalysts the π-complexes with linear alkyl have lower energy. This is true for both ethylene and propylene complexes. A similar effect was observed for both, the diimine and anilinotropone systems (*16-18*).

The presence of the steric bulk also affects the olefin complexation energies, and the relative stability of ethylene and propylene complexes. This has been discussed in details in ref. 17.

Olefin insertion barriers, 1,2- vs. 2,1-propylene insertion

Comparing the systems with isomeric alkyls, the olefin insertion barriers increase from the systems with primary alkyl to those with tertiary alkyl (*16*). As a result, the insertion from the 'tertiary' π-complexes practically does not happen. In the case of the 'real' Pd-diimine systems the TS for the 'secondary' ethylene insertion has slightly higher energy than the 'primary' TS. The computed value of 0.5 kcal/mol is slightly lower then the experimental result (1 kcal/mol, Figure 1).

For the generic catalyst the 2,1-insertion of propylene is preferred by. ca. 2 kcal/mol over the 1,2-insertion. The origin of this preference was discussed in details in ref. 16. The propylene 1,2-insertion TS are hardly affected by the steric bulk (*17*). However, the TS for the propylene 2,1-insertion is strongly destabilized due to a repulsion between the methyl group of propylene and the catalyst substituents (*17*). As a result, in the real systems the 1,2-insertion becomes preferred. As we will discuss later, this is of great importance for the branching control in the propylene polymerization.

Chain isomerization barriers

In the case of Pd-diimine catalyst, the isomerization barriers (5.5 – 8.0 kcal /mol) are substantially lower than the insertion barriers (17.5-18.5 kcal/mol) (*1,16,17,23-25*). The difference between the isomerization and insertion barriers is substantially larger than for the corresponding Ni-complexes, for which the isomerization barriers are higher by c.a. 5-8 kcal/mol and the insertion barriers are lower by 4-5 kcal/mol (2, *26-29*). As we will show later, this is responsible for

160

a difference in the pressure effect on the branching number observed for Ni- and Pd-based systems (*19*).

The energetics of the isomerization reactions is determined by the stability of the alkyl complexes discussed earlier. We would like to point out only that as a result, the isomerization going from primary to secondary carbon is always energetically favored. Also, the 'secondary C $\rightarrow$ secondary C' isomerization reactions have usually lower barriers than the 'secondary C $\rightarrow$ primary C' isomerizations. This implies a fast 'walking' along the polymer chain.

Stochastic simulations of the polyolefin growth

The details of the model has been described in a recent paper (*15*). The stochastic method uses as input data the energies / activation energies of the elementary reactions in the process and is based on the assumption that the relative probablilities of two reactive events (microscopic), π_i and π_j , are equal to their relative reaction rates (macroscopic), $\pi_i / \pi_j = r_i / r_j$;   with the probability normalization constraint, $\Sigma_i \pi_i = 1$. Such an approach makes it possible to model the temperature and pressure effects. The temperature dependence appears in the probabilities for all the events through the rate constants and equilibrium constatnts (Arrhenius / Eyring equations). The pressure affects directly the relative probabilities for unimolecular (isomerization) / bimolecular (ethylene capture + insertion) reactive events, and indirectly it influences all the probabilities because of probability normalization. Namely, the relative isomerization/insertion probability is given by

$$\pi_{iso}/\pi_{ins} = r_{iso} / r_{ins} = k_{iso} / (k_{ins} K p) \tag{1}$$

where k_{iso} and k_{ins} are the insertion and isomerization rate constants, K denotes the ethylene complexation equilibrium constant, and p stands for the pressure.

Ethylene polymerization catalyzed by Pd-diimine catalyst

The simulations (*19*) performed for the ethylene polymerization catalyzed by the Pd-based diimine catalyst **7** (Scheme 1) with the theoretically determined energetics (*16,17*) of elementary reactions gave the average branching number of 131 branches / 1000 C. This is in a reasonable agreement with the experimental value of 122 br./1000 C (*4*) obtained for a closely related system **6**. To further validate our model, we performed the simulations with the experimental values (*23-25*) of the reaction barriers and stabilities of intermediates (see Scheme 1). The value of 121 br. /1000 C was obtained, in a very good agreement with experiment.

Figure 2 presents the temperature and pressure effects on the average number of branches. The results indicate an increase in the number of branches with an increase in temperature, in agreement with experimental observations (*3-6*). This trend can be understood from the energetics of the catalytic cycle (see

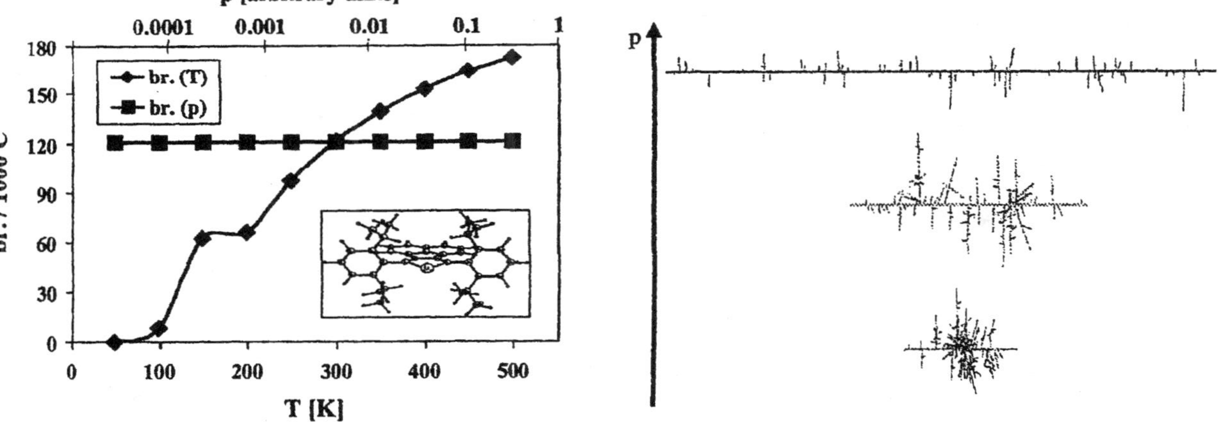

*Figure 2. The pressure and temperature dependence of the average number of branches in the ethylene polymerization catalyzed by the Pd-based diimine catalyst 7 obtained from the stochastic simulations based on experimental energetics (**left**). An influence of pressure on the polymer topology (**right**); in each case a chain of 1000 carbon atoms is shown and the main chain of the polymer is projected along the horizontal axis.*

Figure 1). The TS for secondary insertion is higher in energy (by 1 kcal/mol) than the TS for the primary insertion. Thus, in the limit of T = 0 K only primary insertions may happen, and with an increase in temperature the fraction of the secondary insertions increases.

Figure 2 shows that within the wide range of pressures the simulations give a constant average number of branches. The microstructure of the polymer, however, is strongly affected by the pressure. Examples of the polymer structures obtained from the simulations are shown in Figure 2. The polymers obtained at high pressures are mostly 'linear' with a large fraction of atoms located in the main chain, and with relatively short and mostly linear side-chains. With a decrease in pressure the hyperbranched structures are formed. Both, the pressure independence of the branching number and the pressure influence on the polymer topology are in agreement with experimental data for Pd-catalysts (*2-6*).

Propylene polymerization catalyzed by the Pd-diimine catalysts

For the propylene polymerization catalyzed by the complexes **1-7** (Scheme 2) the simulations were performed (*15*) based on the calculated energetics of the elementary reactions (*16,17*). For system **6**, the calculated average number of branches is 238 br./ 1000 C, which is slightly larger than the experimental value of 213 br./1000 C. However, the temperature and pressure dependence of the number of branches and the polymer microstructure are in-line with experimental observations (*6*): 1) an increase in polymerization temperature leads to a decrease in the number of branches; 2) olefin pressure does not affect the branching number, but affects the topology, leading to hyperbranched structures at low p.

Further, the simulations confirm the experimental interpretation of the mechanistic details of the process with the catalyst **6** (*6*): 1) both 1,2- and 2,1-insertion happen with the ratio of c.a. 7:3 ; 2) there are no insertions at the secondary carbons; 3) most of the 2,1-insertion are followed by a chain straightening isomerization. Thus, for this catalyst the total number of branches is controlled exclusively by the 1,2-/2,1- insertion ratio (*15*).

We would like to emphasize here, that the branching of polypropylene is controlled by different factors than that of polyethylene (*19*). In the case of ethylene the primary/secondary insertion ratio is crucial, whereas in the propylene polymerization catalyzed by diimine catalysts, the ratio between the two alternative insertion pathways (1,2- and 2,1-) is more important (*15*). As a result, an opposite temperature effect has been observed for ethylene (increase in branching number with T) and propylene (decrease in branching number with T).

Ethylene polymerization catalyzed by the Ni-anilinotropone catalyst

The results of the simulations based on the DFT-calculated energetics (*18*) for the Ni-anilinotropone catalyst are presented in Figure 4. The simulations reproduced (*18*) with a reasonable agreement the experimental pressure and temperature dependence of the average number of branches (*12*). The temperature dependence (increase in branching number with an increase in temperature) can be rationalized in a similar way as for diimine catalysts.

In addition, the stochastic simulations provided detailed information about the topology of the polyethylenes produced in these processes (*18*). The results indicate that a variation in pressure affects the polymer topology more than a variation in temperature. At the low pressure regime the length of the branches increases and the branch-on-branch structures are formed more often.

Model simulations – effect of catalyst variation

To model branching features of the ethylene polymerization processes catalyzed by different single-site catalysts we performed a set of model simulations (*19*) in which the barriers for the primary and secondary insertions were systematicaly changed, ΔE_1, $\Delta E_2 = 1 - 9$ kcal/mol (see Figure 1). A change of the two insertion barriers corresponds to a change of the catalyst, as the processes catalyzed by different systems are characterized by different free energy profiles.

Figure 4 presents the pressure dependence of the average number of branches for different sets of the insertion barriers. The curves of Figure 4 can be understood when one notices that (i) in the limit of p→∞ (cf. Eq. 1) the number of branches must be zero for each catalyst, as ethylene π- complexes would be formed immediately after the insertion and there would be no isomerization; (ii) in the limit of p→0 chain walking would be infinitely fast compared to insertions (cf. Eq. 1), and then the number of branches would be controlled exclusively by the ratio of primary vs. secondary insertions (i.e. the insertion /isomerization ratio would not play a role). Thus, in the low pressure range, the curves of Figure 4 converge to the average number of branches characteristic for a given pair of [ΔE_1, ΔE_2], while in the high pressure regime they decay to zero. Consequently, for each catalyst there exist a range of (low) pressure values, for which the average number of branches is pressure independent. With an increase of the insertion barriers this range extends towards high pressure values (see Figure 4).

It may be concluded from the results of Figure 4 that the faster chain walking is compared to insertions, the more extended is the range of pressures for which the average number of branches is constant. This allows us to qualitatively explain the difference between the Pd- and Ni-based diimine catalysts. It is known that for the Pd-diimine systems the isomerization is much faster compared to insertion than for Ni-complexes (*2,26-29*). As a result, for the Pd-systems the number of branches is constant in the experimental range of pressure values, while for the Ni complexes the number of branches depends on the pressure in this range (and would be pressure independent for much lower pressures). It should be pointed out that the above discussion is consistent with the kinetic arguments given by Guan et al. (*3*) for the Pd-diimine catalyst.

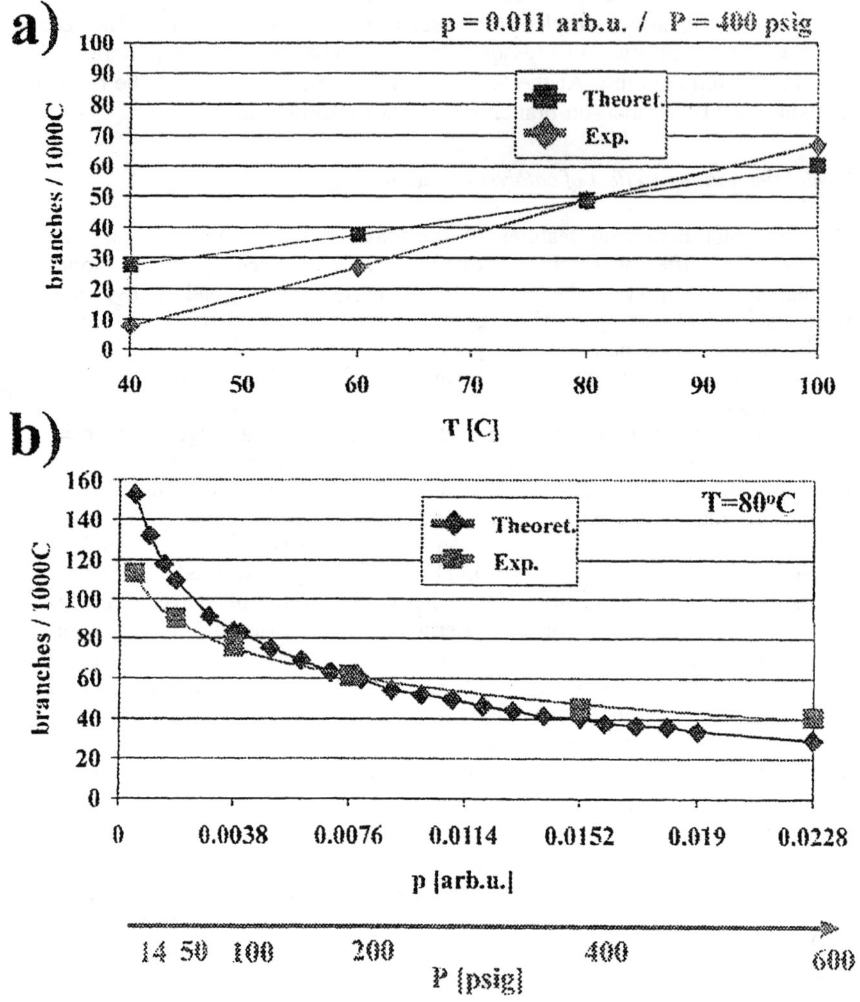

*Figure 3.The calculated and experimental temperature (**a**) and pressure (**b**) dependence of the average number of branches /1000 C.*

Finally, let us now discuss the influence of the pressure on the polymer topology. Figure 4 displays examples of the polymer structures obtained for different combinations of ΔE_1, ΔE_2, for p=0.0001. Although all the structures of Figure 4 are practically hyperbranched, the topology varies to a large extent. As we discussed earlier, in the low pressure regime the branching is controlled by the ratio of primary and secondary insertions. Thus, the difference between the barriers for the primary and secondary insertions controls the polymer topology. For small values of $\Delta E_{2,1} = \Delta E_2 - \Delta E_1$ branches are formed as a result of both, the primary and secondary insertions with comparable probablilities. With an increase in $\Delta E_{2,1}$, the probability of secondary insertions decreases and insertions at the ends of branches happen more often; as a result an increase in the length of linear segments can be observed.

Copolymerization of ethylene with methyl acrylate

An incorporation of polar monomers into a polymer chain by coordination copolymerization is only possible if the polar monomer insertion follows the same reaction mechanism as that of α-olefin Ziegler-Natta polymerization. Thus, before insertion, the polar monomer must be bound to the metal center by its double C=C bond (π-complexes) rather than by the oxygen atom of the polar group (O-complexes). Initially, the mechanism of copolymerization involves a competition between olefin and the polar monomer (Scheme 3). The monomer (polar and non-polar) insertion follows the same mechanism as in the case of homopolymerization (see Scheme 1). However, the most stable products of the polar monomer insertion are complexes involving chelating metal-oxygen bond (Scheme 4). The resting state in the acrylate copolymerization catalyzed by the Pd-diimine catalysts is the 6-member chelate (9). A next step involves an opening of one of the chelates by another monomer molecule, after which the catalytic cycle can repeat itself.

Polar monomer binding mode

The static DFT calculations were performed (20) for the complexes of methyl acrylate, vinyl acetate and their fluorinated derivatives with the Ni- and Pd- diimine and salicylaldiminato catalysts. The results indicate that in the case of the Ni-based diimine system the polar monomers are bound by the carbonyl oxygen atom, while in the Pd-systems the π-complexes are preferred by c.a. 3 kcal/mol. These preferences are practically independent of the steric bulk on the catalysts. The difference between the Ni- and Pd-systems has mainly a steric (electrostatic + Pauli repulsion) origin; there is practically no difference in the orbital-interaction contribution to the binding energy, as far as a comparison between the two binding modes is concerned.

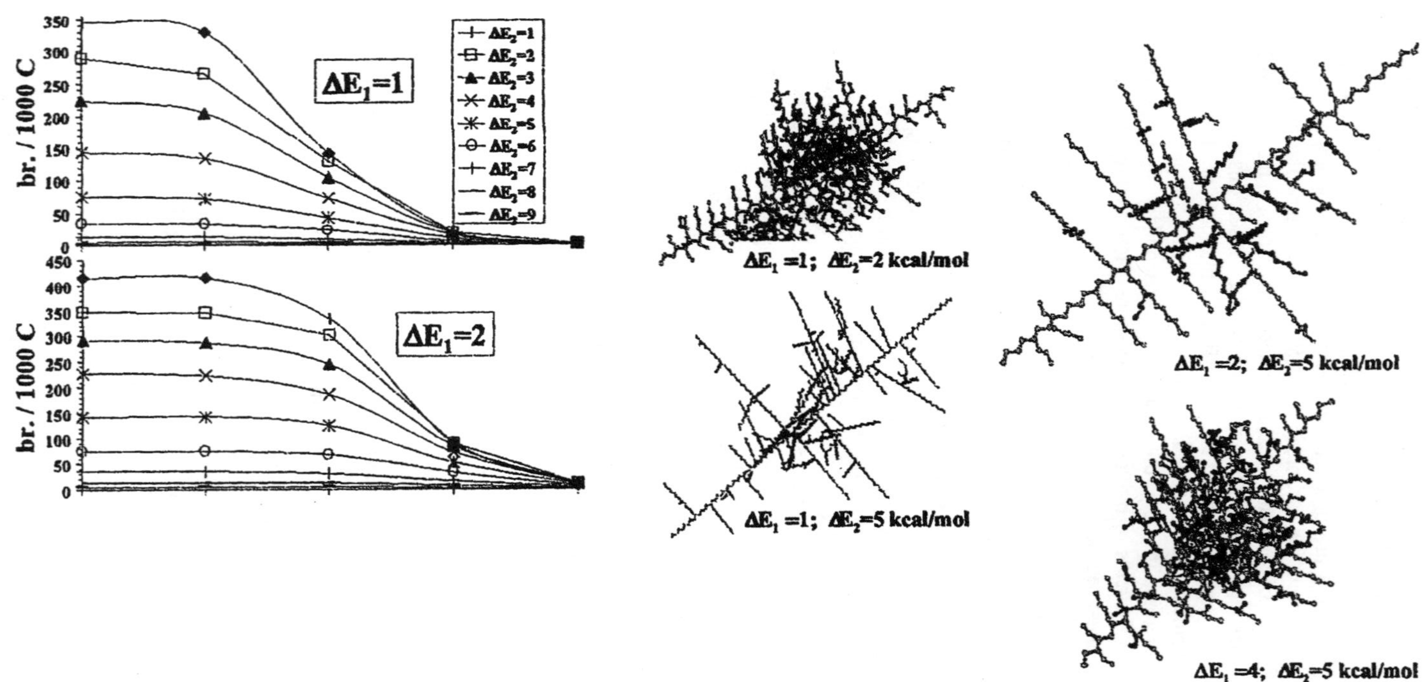

br. / 1000 C
350
300
250
200
150
100
50
0
$\Delta E_1 =1$
$\Delta E_2 =1$
$\Delta E_2 =2$
$\Delta E_2 =3$
$\Delta E_2 =4$
$\Delta E_2 =5$
$\Delta E_2 =6$
$\Delta E_2 =7$
$\Delta E_2 =8$
$\Delta E_2 =9$
br. / 1000 C
450
400
350
300
250
200
150
100
50
0
$\Delta E_1 =2$
$\Delta E_1 =1; \ \Delta E_2 =2$ kcal/mol
$\Delta E_1 =1; \ \Delta E_2 =5$ kcal/mol
$\Delta E_1 =2; \ \Delta E_2 =5$ kcal/mol
$\Delta E_1 =4; \ \Delta E_2 =5$ kcal/mol

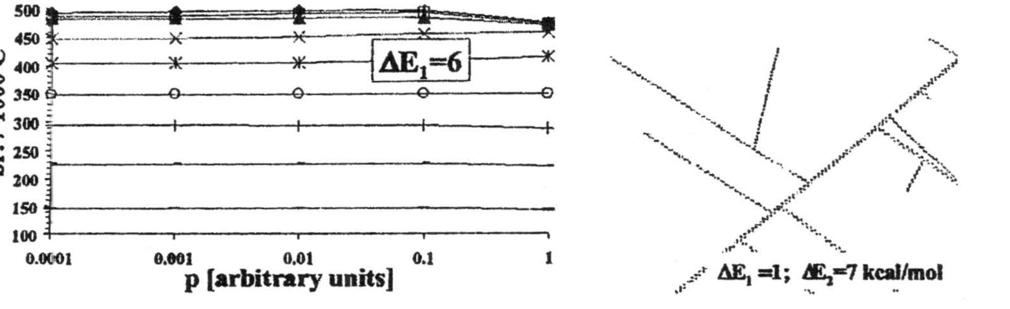

*Figure 4. The pressure dependence of the average number of branches obtained from the model stochastic simulations with different primary and secondary insertion barriers, ΔE_1, ΔE_2 **(left)** and examples of polymer topologies **(right)**. Different atom shadings are used to mark different types of branches (primary, secondary, etc.).*

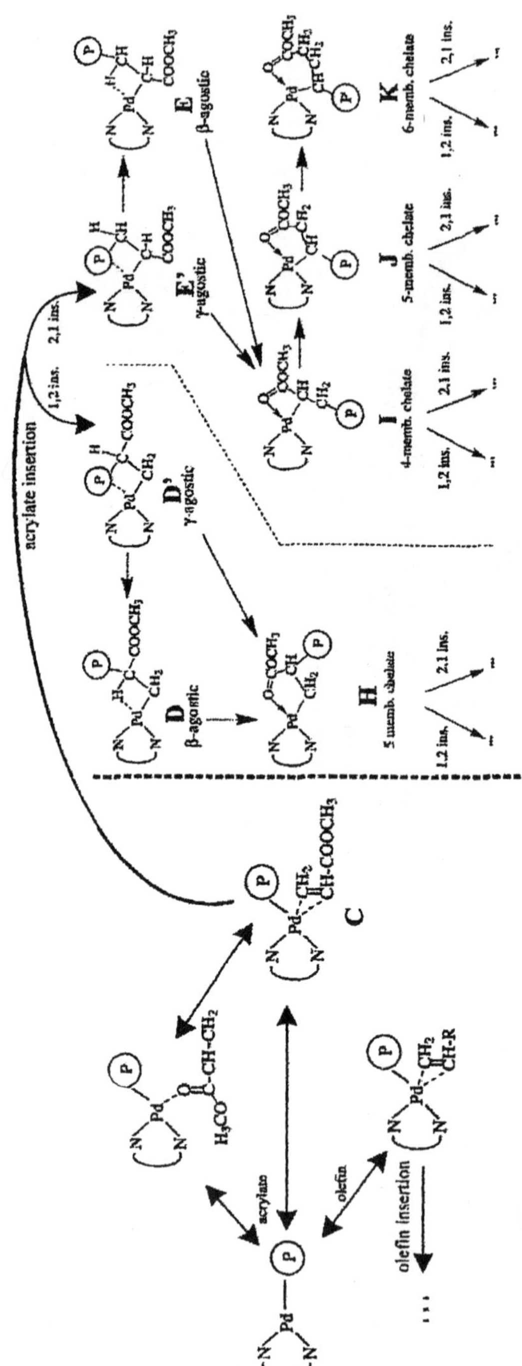

*Scheme 3. Mechanism of the copolymerization of α-olefins with polar monomers. Initial steps involve a competition between the non-polar and polar olefin and its two binding modes (**top**). The polar monomer insertion is followed by formation of chelates (**bottom**).*

In the case of the fluorinated polar monomers, the complexation energies of both, the π- and O-complexes are decreased in comparison to methyl acrylate and vinyl acetate. Thus, in a prospective co-polymerization with fluorinated compounds, the incorporation of the polar monomer into a polyolefin chain would be relatively modest.

Use of neutral catalysts seems to be more promising. The results for Ni- and Pd-based systems with Grubbs ligands show that for both metals the π-complexes are strongly preferred over the O-complexes (by c.a. 6 kcal/mol); the binding energies of the π-complexes are comparable with the corresponding systems involving the diimine catalysts (20). The DFT studies on the methyl acrylate salicylaldiminato systems: the 'real' π-complexes are more stable by c.a. 8-9 kcal/mol than the O-complexes (18).

A comparison of Ni- and Pd-diimine systems (inactive and active catalysts, respectively) leads to the conclusion that the theoretical analysis of the polar co-monomer binding mode can be used as a screening test to select the best prospective catalytic candidates for the co-polymerization: the complexes with preference of the O-complexes can be excluded from further studies.

Finally, it should be mentioned that similar studies were performed for the nitrogen-containing monomers (30). They lead to the same conclusions about the role of the electrophilicity on the metal atom.

Methyl acrylate insertion and the chelate opening

Figure 4 presents the energy profiles for the methyl acrylate insertion into the Pd-alkyl and Ni-alkyl bond in the generic diimine systems (21,22). The results clearly indicate that it is not the insertion barrier which makes the Ni-system inactive in polar copolymerization. The acrylate insertion barrier is substantially lower for the Ni-catalyst then for the Pd-complex. This, in fact, should not be surprising, as the ethylene insertion barriers are also lower for the Ni-system.

The results of Figure 4 also show that the chelate complexes are more stable for the Ni-catalyst then for the Pd-complexes. This larger stability is again a manifestation of the stronger oxophilicity of the Ni-systems.

The static and dynamic DFT studies (21,22) on the chelate opening reactions reveal that the two-step mechanism of the chelate opening must be assumed in order to explain the difference between the Ni- and Pd-based systems. Namely, in the most stable ethylene complexes resulting from the chelates, the chelating metal-oxygen bond is still present in the axial position. The ethylene insertions starting from such structures have very high barriers (ca. 30 kcal/mol for the 6-membered chelate with the generic Pd-catalyst). Much lower barriers, comparable to those of ethylene homo-polymerization (c.a. 18 kcal/mol for a generic Pd-catalyst) have been obtained for the insertions starting from the higher energy isomers in which the chelating bond has been broken. In both cases the insertion barriers computed for the Ni-catalyst are lower than for the Pd-system. However, the opening of the chelate prior to ethylene insertion has substantially lower barrier for Pd- (c.a. 9 kcal/mol) than for Ni-catalyst (c.a 15-19 kcal/mol).

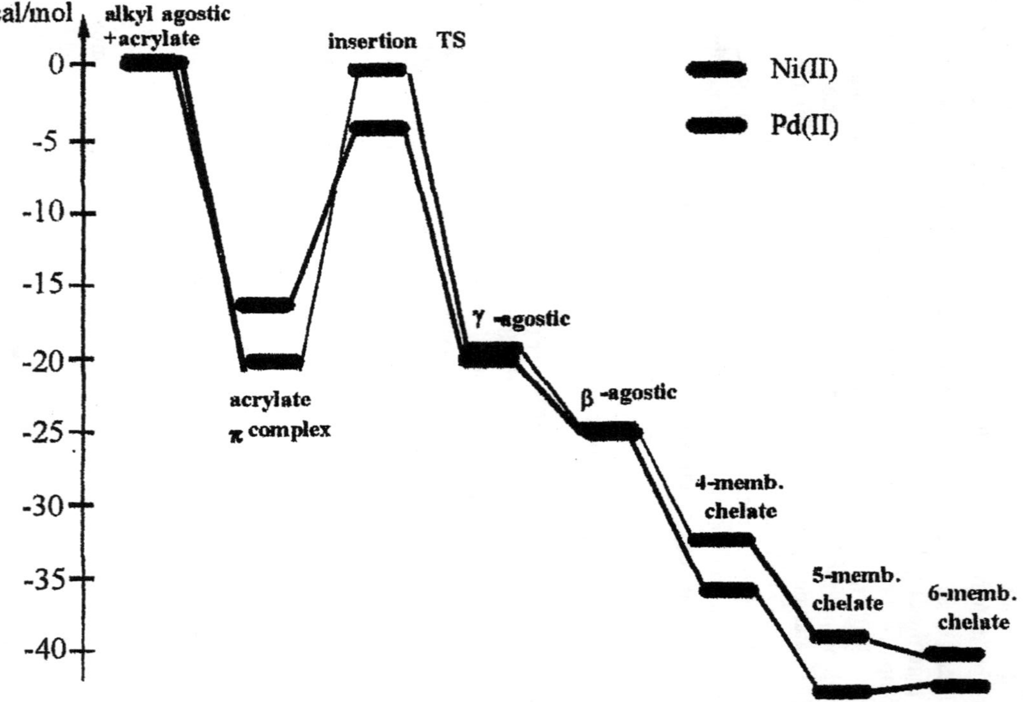

Figure 4 Energy profile for 2,1-insertion of methyl acrylate into a Pd(II)-alkyl bond (black) or a Ni(II)-alkyl bond (gray).

Concluding remarks

We have presented a comprehensive review of our recent theoretical studies on the polymerization and copolymerization processes catalyzed by the late transition metal complexes. The results of these studies show that a combined DFT/stochastic approach can be successfully used to model the elementary reactions in the polymerization processes and the influence of the reaction conditions on the polyolefin branching. Such an approach makes it possible to understand the microscopic factors controlling the branching of polyolefins and explain the differences between the Pd- and Ni-catalysts. The results also demonstrate that a wide range of microstuctures can be potentially obtained from the ethylene polymerization. Thus, a rational design of the catalyst producing the desired polymer topology should be possible.

The result of the static and dynamic DFT calculations for the methyl acrylate copolymerization suggest that there are two factors inhibiting the polar co-polymerization in the Ni-catalyst case: (1) the initial O-complex formation; (2) a difficult chelate opening prior to insertion of the next monomer. Both of those factors may be overcome by the use of the complexes with reduced oxophilicity of the metal on the catalyst.

References

1. Britovsek, G.J.P.; Gibson, V.C.; Wass, D.F. *Angew. Chem. Int. Ed.* **1999**, 38, 428, and refs therein.
2. Ittel, S.D.; Johnson, L.K.; Brookhart, M. *Chem. Rev.* **2000,** 100, 1169, and refs. therein.
3. Guan, Z.; Cotts, P.M.; McCord, E.F. ; McLain, S.J. *Science,* **1999,** 283, 2059.
4. Cotts, P.M.; Guan, Z.; McCord, E.F.; McLain, S.J. *Macromolecules* **2000,** 33, 6945.
5. Gates, S.J. *et al.*, *Macromolecules* **2000**, 33, 2320.
6. McCord, S.J. *et al.*, *Macromolecules* **2001**, 34, 362.
7. Boffa, L.S.; Novak, B.M. *Chem. Rev.* **2000,** 100, 1479, and refs. therein.
8. Johnson, L. K.; Mecking, S.; Brookhart, M. *J. Am. Chem. Soc.* **1996**, *118*, 267.
9. Mecking, S.; Johnson, L. K.; Wang, L.; Brookhart, M. *J. Am. Chem. Soc.* **1998**, *120*, 888.
10. Johnson, L.K.; Killian, C.M.; Brookhart, M. *J. Am. Chem. Soc.* **1995**, 117, 6414.
11. Killian, C.M.; Tempel, D.J.; Johnson, L.K.; Brookhart, M. *J. Am. Chem. Soc.* **1996**, 118, 11664.
12. Hicks, F.A.; Brookhart, M. *Organometallics* **2001**, 20, 3217.
13. Rappe, A.K.; Skiff, W.M.; Casewit, C.J. *Chem. Rev.* **2000**, 100, 1435, and refs. therein.
14. Angermund, K.; Fink, G.; Jensen, V.R.; Kleinschmidt, R. *Chem.Rev.* **2000**, 100, 1457, and refs. therein.
15. Michalak, A.; Ziegler, T. *J. Am. Chem. Soc.* **2002**, 124, 7519.
16. Michalak, A.; Ziegler, T. *Organometallics* **1999**, 18, 3998.
17. Michalak, A.; Ziegler, T. *Organometallics* **2000,** 19, 1850.

18. Michalak, A.; Ziegler, T. *submitted.*
19. Michalak, A.; Ziegler, T. *submitted.*
20. Michalak, A.; Ziegler, T. *Organometallics,* **2001,** 20, 1521.
21. Michalak, A.; Ziegler, T. *J. Am. Chem. Soc.* **2001**, 123, 12266.
22. Michalak, A.; Ziegler, T. *unpublished work.*
23. Tempel, D.J. ; Johnson, L.K.; Huff, R. L.; White, P.S. ; Brookhart, M. *J. Am. Chem. Soc.* **2000**, 122, 6686.
24. Shultz, L.H.; Brookhart, M. *Organometallics,* **2001,** 20, 3975.
25. Shultz, L.H.; Tempel, D.J.; Brookhart, M. *J. Am. Chem. Soc.* **2001,** 123, 11539.
26. Musaev, D.G; Froese, R.D.J.; Morokuma, K. *Organometallics* **1998,** 17, 1850.
27. Froese, R.D.J; Musaev, D.G.; Morokuma, K. *J. Am. Chem. Soc.* **1998**, 120, 1581.
28. Deng, L.; Margl, P.; Ziegler, T. *J. Am. Chem. Soc.* **1997**, 119, 1094.
29. Deng, L.; Woo, T.K.; Cavallo, L.; Margl, P.M.; Ziegler, T. *J. Am. Chem. Soc.,* **1997**, 119, 6177.
30. Deubel, D.; Ziegler, T. *Organometallics,* **2002,** 21, .

Acknowledgment This work has been supported by the National Sciences and Engineering Research Council of Canada (NSERC) as well as donors of the Petroleum Research Fund, administered by the American Chemical Society (ACS-PRF No. 36543-AC3). AM acknowledges the NATO Postdoctoral Fellowship. TZ thanks the Canadian Government for a Canada Research Chair. Important part of the calculations was performed using the UfC MACI cluster.

Chapter 13

Vinyl-Type Polymerization of Norbornene by Nickel(II) bis-Benzimidazole Catalysts

Abhimanyu O. Patil, Stephen Zushma, Robert T. Stibrany, Steven P. Rucker, and Louise M. Wheeler

Corporate Strategic Research, ExxonMobil Research and Engineering Company, 1545 Route 22 East, Annandale, NJ 08801

Novel nickel (II) bis-benzimidazole complexes were prepared. These complexes were synthesized using three steps: condensation, alkylation and complexation. These complexes, upon activation with MAO, polymerize cyclic olefin norbornene. The polynorbornene synthesized was characterized by GPC, ^{13}C NMR and IR. GPC molecular weights of the polynorbornene polymers were determined using the LS/viscometry/DRI triple detector system and were found to be very high (Mn 148,000 to 288,000 and Mw 587,000 to 797,000). The ^{13}C NMR suggest that the polynorbornene is a vinyl addition product (*i. e.,* 2,3-linked); no ring-opened product was observed.

Late transition metal catalyst complexes are becoming important for oligomerization and polymerization of olefins (*1,2*). Previously, we have reported the synthesis of copper (II) bis-benzimidazole complexes (*3*). Cu-complexes, on activation with MAO, polymerize ethylene to highly linear polyethylene. Similar to copper (II) bis-benzimidazole complexes, nickel (II) bis-benzimidazole complexes can be prepared. Both types of complexes crystallize in pseudo-tetrahedral geometry. Nickel (II) bis-benzimidazole complexes, on activation with MAO polymerize norbornene.

Polynorbornenes are of considerable interest because of their unique physical properties, such as high glass transition temperature, optical transparency, and low birefringence. Bicyclo[2,2,1]hept-2-ene, i.e., norbornene, and its derivatives can be polymerized in several different ways (Scheme 1). It is noteworthy that each route leads to its own polymer type that is different in structure and properties.

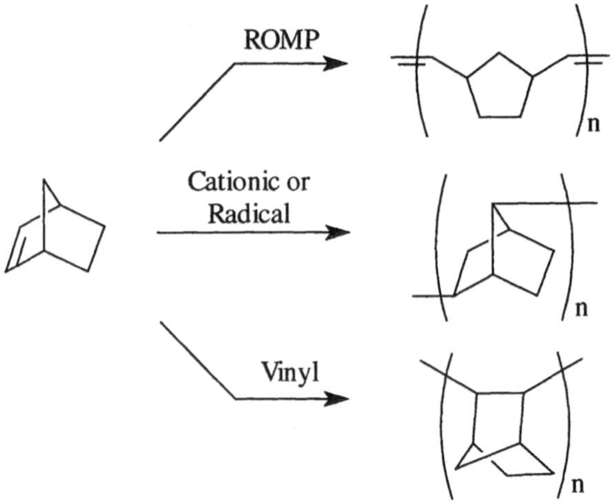

Scheme 1: Schematic representation of the three different types of norbornene polymerizations.

The polymerization of norbornene has been known since the 1950's when polymers were prepared via the ring-opening metathesis polymerization (ROMP) of norbornene (*4*). ROMP yields poly(1,3-cyclopentylenevinylene), which retains one double bond in each polymeric repeat unit. This approach has been the most explored among the three approaches (*5*). Since polymer obtained by the ROMP process still contains double bonds in the polymer backbone, the polynorbornene obtained by this process can be vulcanized and is used as an elastomeric material. The commercial polymerization process uses a $RuCl_3/HCl$ catalyst in butanol.

Norbornene can also be polymerized by carbocationic and free radical initiated reactions. Cationic polymerization involves rearrangement of the norbornene framework and generally produces moderate yields of poly(2,7-bicyclo[2,2,1]hept-2-ene) oligomers. Free radical processes also produce low molecular weight products. The initiators for the radical polymerizations were 2,2'-azobisisobutyronitrile (AIBN) or *t*-butyl perpivalate, while $EtAlCl_2$ was used for cationic polymerization (*6-9*).

It is also possible to polymerize norbornene and to leave the bicyclic structural unit intact, i.e., to open only the double bond π-component. Such polymerization is called "vinyl-type" or addition polymerization. The product, poly(2,3-bicyclo[2,2,1]hept-2-ene), a saturated polymer does not contain any double bonds. Hoechst and Mitsui have jointly developed highly transparent plastics based on this approach, thermoplastic olefin polymers of amorphous structure (TOPAS). TOPAS is a norbornene ethylene copolymer prepared via metallocene catalyst (*10-11*).

Thus, the polymerization of norbornene can lead to different polymers depending on the catalyst and mechanism. Metal catalysts in higher oxidation states will form an unsaturated elastomer by ring opening metathesis polymerization (ROMP). On the other hand, a vinyl-type polymer is obtained by addition polymerization of metal catalysts in lower oxidation states. The polymers are crystalline or amorphous depending on the catalyst. Complexes of nickel (*12-13*), titanium (*14*), zirconium (*15-16*), and palladium (*17-19*) have been utilized for vinyl-type polymerization of norbornene.

The focus of this paper is the synthesis of Ni (II) bis-benzimidazole complexes and the polymerization of norbornene with MAO activated Ni (II) bis-benzimidazole complexes. The polynorbornenes synthesized were characterized by GPC, ^{13}C NMR and IR. GPC molecular weights of the polynorbornene polymers were determined using the LS/viscometry/DRI triple detector system and were found to be very high (Mn 148,000 to 288,000 and Mw 587,000 to 797,000). The ^{13}C NMR suggest that the polynorbornene is a vinyl addition product (*i. e.*, 2,3-linked) in that no ring-opened product was observed.

Results and Discussion

Nickel (II) Bis-benzimidazole Complexes

Previously, we have reported the synthesis of copper (II) bis-benzimidazole complexes (*3*). Two analogous nickel (II) bis-benzimidazole complexes with methylene bridged (aliphatic bridge complex **I**) bis-benzimidazole and biphenyl bridged (aromatic bridge complex **II**) bis-benzimidazole are prepared.

Catalyst Synthesis

The nickel complexes are prepared in three synthetic steps. The first step is the condensation reaction of a dicarboxylic acid with two equivalents of phenylenediamine in polyphosphoric acid adapted from a literature procedure (*20*).

Molten polyphosphoric acid (> 160 °C) was used as a solvent and dehydrating agent in this reaction. Condensation of malonic acid with phenylenediamine yields the methylene-bridged (aliphatic bridge) bis-benzimidazole. Condensation of diphenic acid with phenylenediamine yields the 2,2' biphenyl bridged (aromatic bridge) bis-benzimidazole. The products obtained in condensation reactions are high yield products (60-90%).

Alkylation Reaction

The next synthetic step is alkylation of the bis-benzimidazole. Alkylation was carried out using sodium hydride as the deprotonating agent followed by addition of the appropriate iodoalkane. The alkylation is rather complicated by the fact that more than one alkylated product is possible. The methylene bridged bis-benzimidazole can be alkylated sequentially up to four times, once each at the protonated nitrogens and twice at the methylene bridgehead carbon. In this study methylene bridged bis-benzimidazole was alkylated three times. The 2,2' biphenyl bridged bis-benzimidazole was alkylated twice, once at each protonated nitrogen. It was found that the yields of the alkylated products are dependent upon reaction conditions, such as temperature, nature of the base, the dryness of the sample, etc. (*3*).

Complex Formation

Finally, the alkylated bis-benzimidazole compounds were metalated using $NiCl_2 \cdot 6H_2O$ in a mixture of ethanol and triethylorthoformate. Triethylorthoformate is added to effect the dehydration of the reaction mixture and aid in the crystallization of the metal complex. The solid crystalline complexes are then isolated by filtration.

POLYMERIZATION OF NORBORNENE

The polymerization of norbornene can lead to polymers with different properties depending on the type of catalyst and the mechanism of

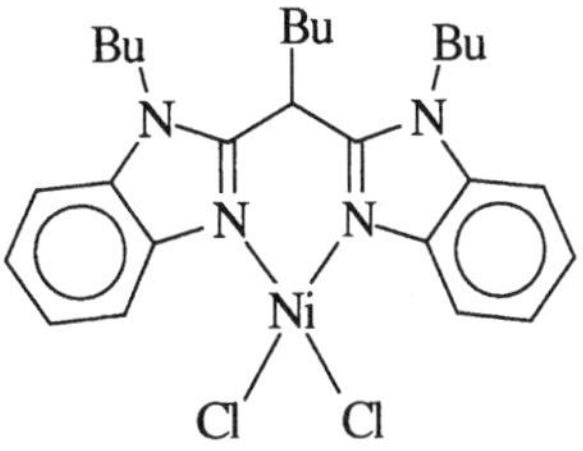

$Ni(Bu_3BBIM)Cl_2$ (**I**)

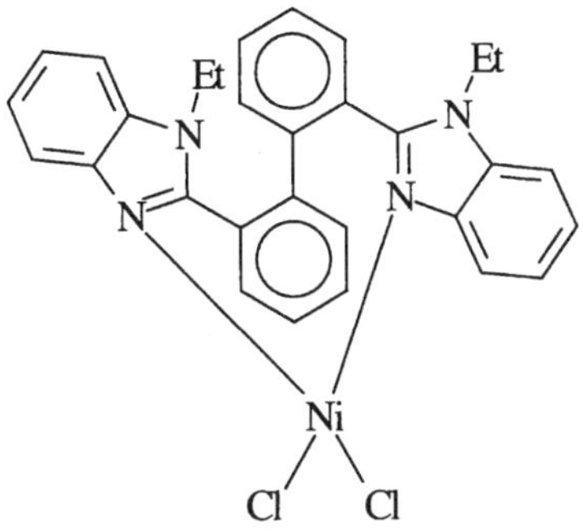

$Ni(Et_2BBIL)Cl_2$ (**II**)

Condensation Reaction

Aliphatic Bridge

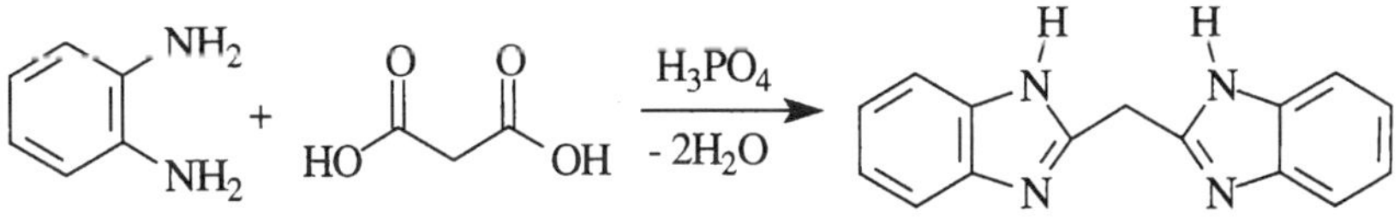

Aromatic Bridge

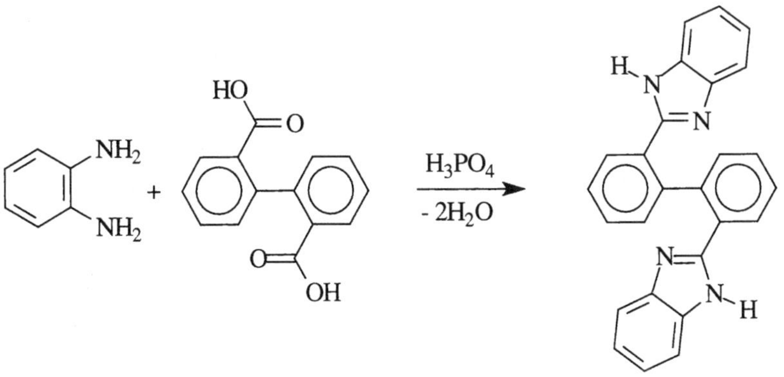

Aliphatic Bridge

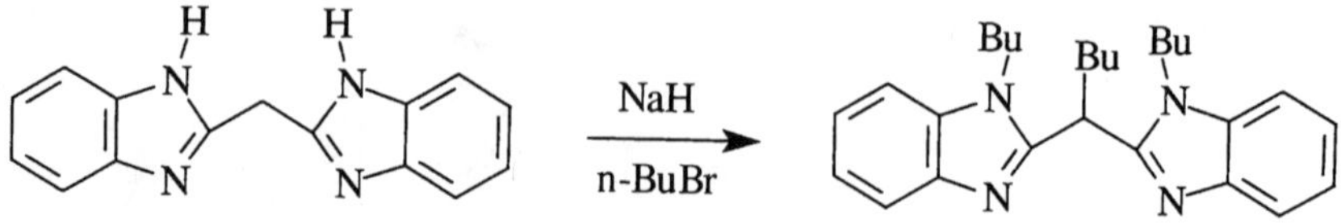

Aromatic Bridge

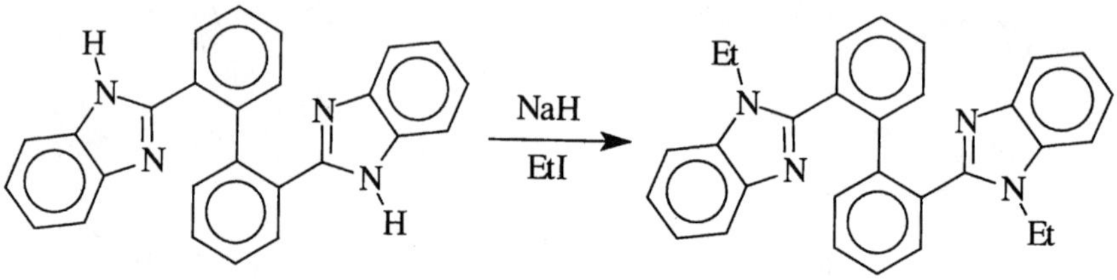

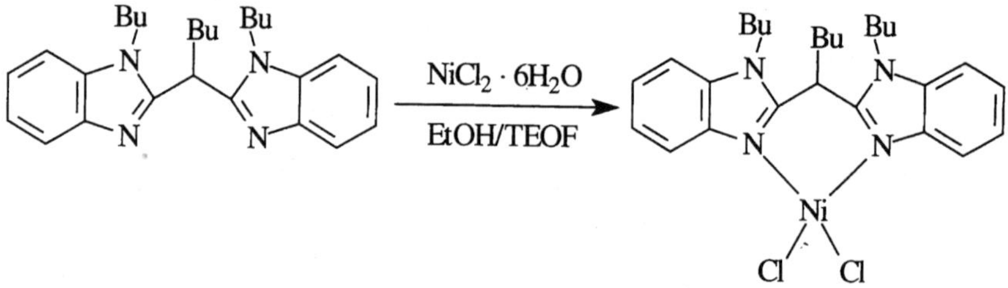

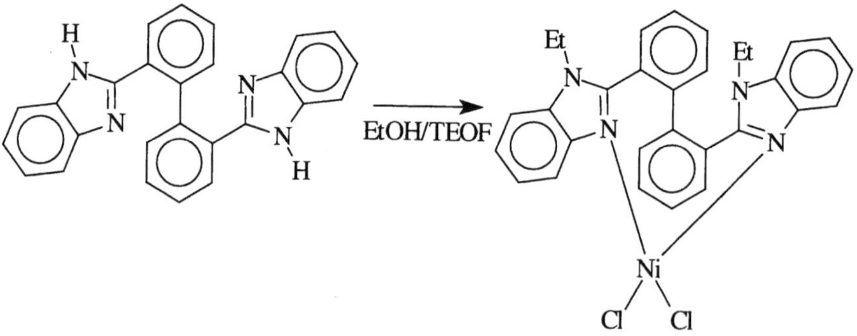

polymerization. There are several reports on the use of late transition metal based complexes as catalysts for "vinyl-type" polymerization of norbornene (*12-13,17-19*). The vinyl homopolymerization of norbornene can be carried out using early transition metals, late transition metals, and rare-earth metals. The polymers are crystalline or amorphous depending on the catalyst. From the materials point of view amorphous polymers are of interest because they are soluble in organic solvents, which allows product analysis and processing. The remarkable differences in solubility characteristic of polynorbornene polymers prepared using different catalysts shows that there is clearly a large difference in polymer microstructure.

Goodall *et al.* (*13*) have reviewed the development of single-component cationic nickel and palladium catalysts and their use in polymerization of norbornenes at high rates, to obtain amorphous, transparent polymers with very good heat resistance. Using NMR methods, the initiation step has been shown to occur exclusively via addition to the *exo*-face of norbornene (*13*). The result of this growth process is a very high molecular weight polynorbornene since the "normal" mode of chain transfer (β-hydride elimination) is not possible given the geometry of the active, growing center. Specifically there are two β-hydrides but neither of them can be eliminated to the metal center because one of the two is located at a bridgehead (and is therefore forbidden, "Bredt's Rule") and the other is *anti* or *"trans"* to the metal. Thus the polymerization is essentially "living" since there is no pathway for chain transfer.

If one compares the Cu and Ni bis-benzimidazole complexes, both types crystallize in pseudo-tetrahedral geometry. Representative Cu and Ni complexes with their bond-angles are shown in Figure 1.

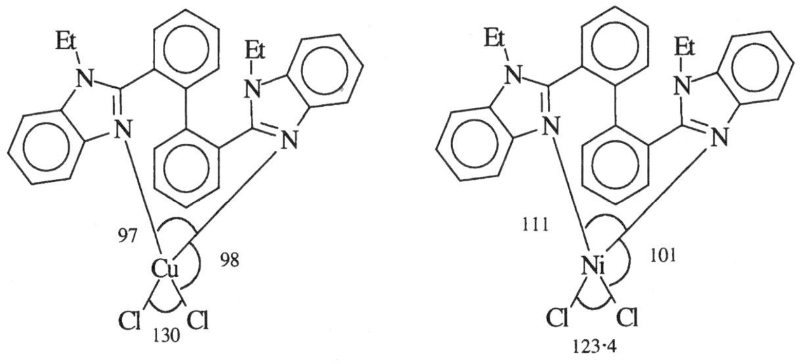

Figure 1. Representative Cu and Ni bis-benzimidazole complexes: Both complexes show pseudotetrahedral geometry.

Polymerization of Norbornene using Ni Bis-benzimidazole Catalysts

The polymerization of norbornene using Ni bis-benzimidazole was studied under a range of polymerization conditions.

Effect of Al/Ni ratio on Norbornene Polymerization

One variable that was examined was the ratio of activator or co-catalyst to metal complex. The norbornene polymerization data using the MAO activated 1,1'-bis(1-butylbenzimidazole-2yl)pentane)nickel(II)dichloride

$[Ni(Bu_3BBIM)Cl_2]$ complex are shown in Table 1. The ratio of monomer to metal complex, reaction time and temperature was kept constant for all three polymerization reactions. The activator to pre-catalyst complex mole ratio (Al/Ni) was varied from 100 to 1500. The amount of precatalyst used was 8.0 mg (1.0×10^{-2} mmol) and reaction temperature was 25 °C. Overall the polymerization reaction reaches high conversion even with an Al/Ni mole ratio of 100.

Table 1. Effect of Al/Ni ratio on norbornene polymerization

Run #	Al/Ni (mole)	Yield (g)	Conv (%)	M_n	M_w
111A	100	0.96	93.7	388,000	797,000
111B	700	1.01	98.4	260,000	776,000
111C	1500	1.07	100.00	297,000	587,000

NB/Ni (mole ratio): 700; Reaction Time: 4 h, Reaction Temp: 25 °C.

The polymers obtained were characterized by ^{13}C NMR, IR and GPC analysis. The ^{13}C NMR spectra of the products were recorded in CDCl$_3$. The NMR spectrum gave a series of resonances δ = 30.7-51.6 (m, maxima at 29.7, 31.5, 35.4, 37.4, 38.5, 39.4, 47.5, 51.6) ppm. A typical ^{13}C NMR spectrum of the product is shown in Figure 2. Greiner *et al.* (*12*) have reported a similar NMR for their vinyl-type norbornene polymer. [^{13}C NMR (CDCl$_3$): δ = 30.7-51.6 (m, maxima at 30.7, 32.4, 36.3, 38.3, 39.4, 40.4, 48.4, 51.6) ppm. These data suggest that the polynorbornene obtained using the above catalyst is a vinyl-type (2,3-linked) addition product; no ring opening polymerization was observed. The IR spectra of the products showed characteristic absorption peaks at 2950 s, 2860 s,

1474 s, 1452 s, 1375 m, 1294 s, 1258 s, 1221 s, 1146 m, 11010 m, 1041 m, 941 m, 891 m, 756 w cm^{-1}. The IR spectrum is identical with the reported IR (*14*).Bands at 730 and 960 cm^{-1}, attributed to the double bonds in cis and trans forms following ring-opening polymerization, are not observed (*14*).

GPC molecular weights of the polynorbornene polymers were determined using the LS/viscometry/DRI triple detector system. The molecular weights of these polymers are very high indicating that the "normal" mode of chain-transfer (β-hydride elimination) is not possible in polynorbornene given the geometry of the active growing center.

Effect of reaction time on Norbornene Polymerization

The polymerization of norbornene using Ni bis-benzimidazole was studied using varied reaction times. Table 2 summarize the norbornene polymerization data using the MAO activated 1,1'-bis(1-butylbenzimidazole-2yl)pentane)nickel (II) dichloride [Ni(Bu$_3$BBIM)Cl$_2$] complex using various reaction times. The ratio of monomer to metal complex, reaction temperature and the activator to pre-catalyst complex mole ratio (Al/Ni) were kept constant in all polymerization reactions. The reaction time was varied from 0.5 h to 4.0 h. The amount of catalyst used was 8.0 mg (1.0 x 10^{-2} mmol) and reaction temperature was 25 ^{o}C. Again the overall polymerization reaction reaches high conversion even within 0.5 h. The reaction is essentially complete in 2 hours.

GPC Analysis of Polynorbornene Samples

GPC analysis of the norbornene polymers was conducted at 135°C using trichlorobenzene as the mobile phase and operating conditions for routine PE and EP analysis. The data were collected on the light scattering (LS) / Viscometry / differential refractive index (DRI) triple detector system. For all the polynorbornene samples, the response from all three detectors was strong with good signal to noise. Figure 3 shows the chromatograms of 115B PNB sample.

For routine polymers such as polyethylene, these polymer parameters are well known and the calculations straightforward. However, because of the novelty of these samples, the dn/dc and Mark-Houwink parameters were not available and had to be estimated from the GPC data. The estimates were done as follows: The first sample, 111A, was initially analyzed using the polymer parameters for polyethylene. The correct dn/dc for polynorbornene was then estimated by comparing the "calculated mass" (from the integrated peak area and PE dn/dc) to the "inject mass" (from the solution concentration and injection volume) and changing the input dn/dc so that the two mass calculations match. The accuracy of this estimation depends on the complete recovery of the

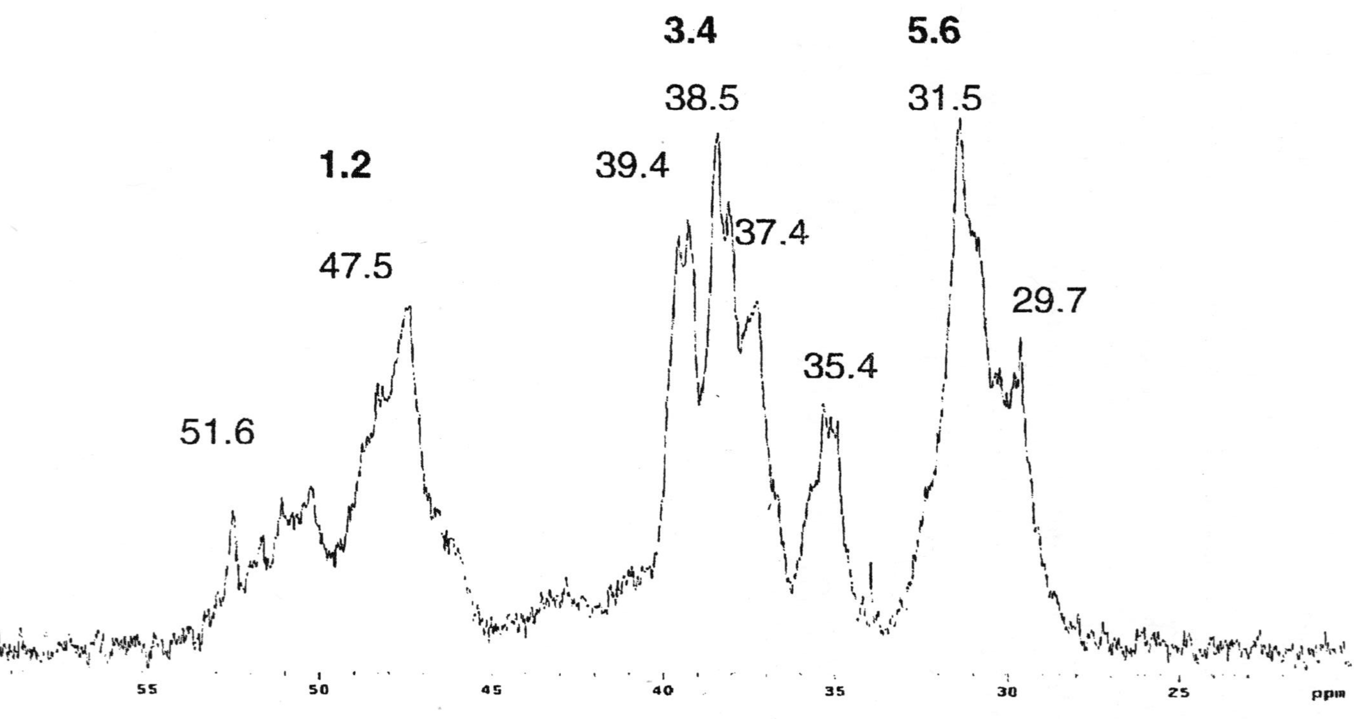

Figure 2. ^{13}C NMR spectrum of polynorbornene prepared by Ni bis-benzimidazole catalyst.

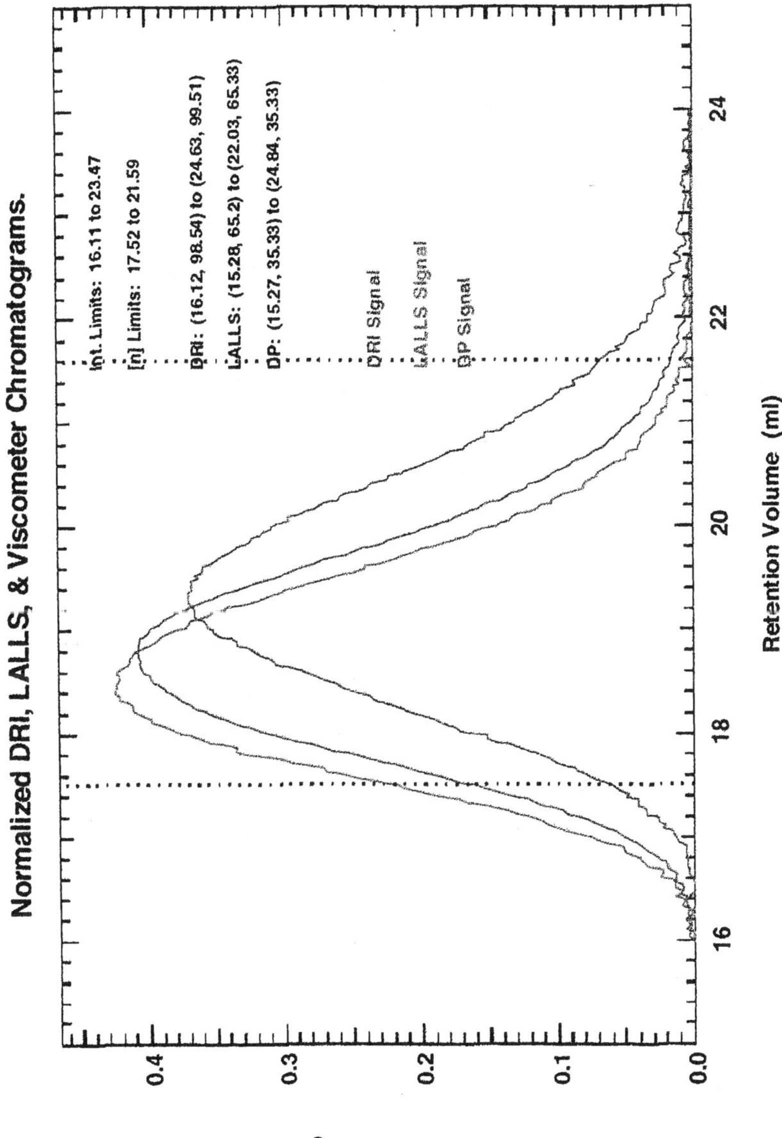

Figure 3. GPC triple detector chromatogram of PNB sample 115B.

Table 2. Effect of reaction time on norbornene polymerization

Run #	Reaction time (h)	Yield (g)	Conv (%)	M_n	M_w
115A	0.5	0.82	81.7	204,000	556,000
115B	1.0	0.94	90.7	278,000	657,000
115C	2.0	1.00	98.9	223,000	592,000
115D	3.0	1.00	98.8	148,000	529,000
111B	4.0	1.01	98.4	260,000	776,000

NB/Ni (mole ratio): 700; Al/Ni (mole%) 700, Reaction Temp: 25 °C.

polymer, i.e. none of the sample was filtered out prior to injection or lost in the columns. Once the dn/dc was determined, the concentration profile, molecular weight averages, and intrinsic viscosities by triple detector could be calculated. The Mark-Houwink parameters were estimated in the analysis program from a fit of the log intrinsic viscosity – log $M_{triple\ detector}$ plot for sample 111A. The accuracy of this estimation is dependent on the linearity of the sample. If the sample contains long chain branching, the intrinsic viscosity will be depressed compared to a linear analogue of equivalent molecular weight and the Mark-Houwink fit will be incorrect. If the fit is good, the k and alpha can be applied to the entire set of polymers having the same composition (above a minimum molecular weight) with reasonable results.

The results of these calculations for polynorbornene yielded a dn/dc of 0.046, and Mark-Houwink k and alpha values of 0.6793 and $9.872e^{-5}$, respectively. Again, it should be noted that these results were based only on the data for the first sample, 22793-111A, and that calculations based on any other particular polynorbornene sample may yield slightly different but comparable polymer parameters. The validity of this approach is substantiated by comparing the results in the Table 3. The calc/inj mass ratio for the pure polynorbornenes agree within 10%, except for 22793-115D which had a recovery of only 82% based on a dn/dc of 0.046. The molecular weights calculated by triple detector and by DRI alone show consistency if not absolute agreement. Finally, the g' calculations are all very close to 1.0 and the log intrinsic viscosity – log M plots are all fairly straight. An example of the Mark-Houwink plot and g' branching distribution is shown in Figure 4.

TGA analysis of Polynorbornene

One of the polynorbornene samples (111B) was evaluated by thermogravimetric analysis (TGA). The polymer sample was heated at 10 °C/min in a nitrogen atmosphere. The TGA scan is shown in Figure 5. The polymer is very stable up to 400 °C.

Intrinsic Viscosity vs. Molecular Weight by Triple Detection.

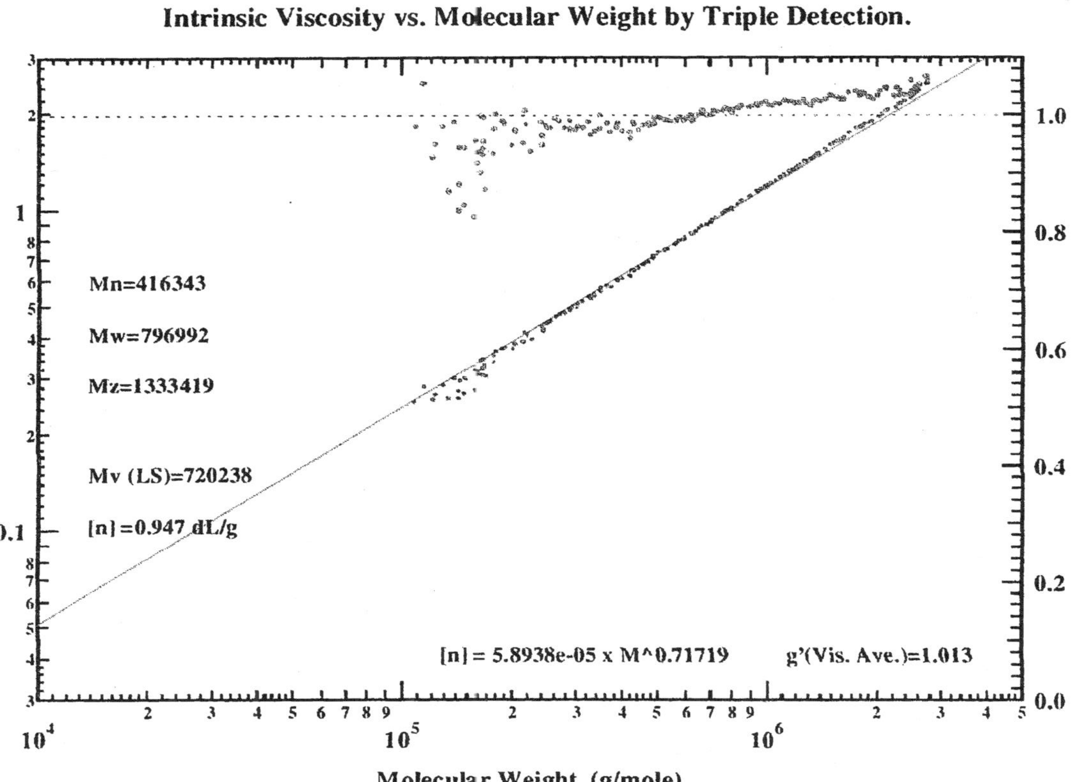

Figure 4. Mark-Houwink plot and branching distribution for PNB sample 115B.

186

Table 3. GPC Results for polynorbornene.

Run #	Calc/Inj Mass	M_n Triple Det	M_w Triple Det	Average g'	M_n DRI	M_w DRI
111A	1.00	388,000	797,000	1.01	347,000	869,000
111B	1.07	260,000	776,000	0.99	221,000	819,000
111C	0.92	297,000	587,000	0.99	268,000	633,000
115A	1.01	204,000	556,000	1.01	193,000	587,000
115B	1.01	278,000	657,000	1.05	274,000	740,000
115C	1.05	233,000	592,000	1.01	210,000	678,000
115D	0.82	148,000	529,000	0.98	148,000	612,000

Dn/dc = 0.0480

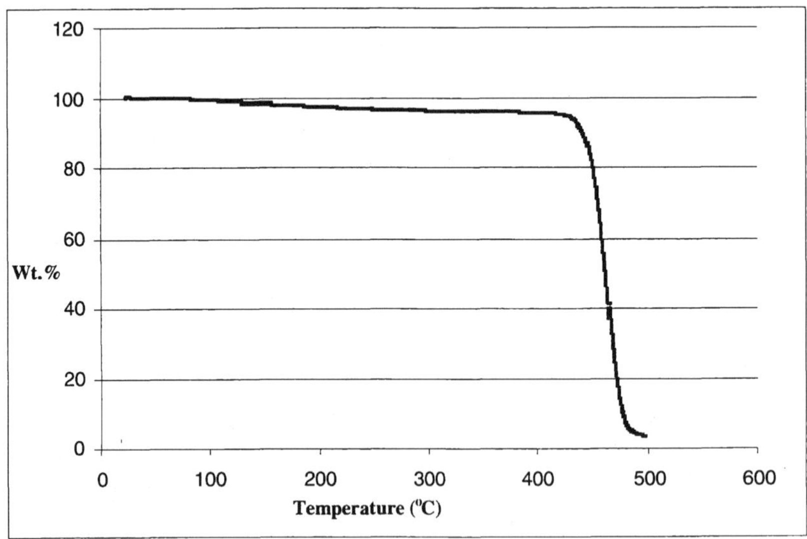

Figure 5. TGA of polynorbornene homopolymer.

EXPERIMENTAL SECTION

Catalyst Syntheses

General Alkylations utilizing NaH were conducted under nitrogen using Schlenk techniques. Dimethyl sulfoxide was dried over CaH_2. Toluene and THF were purified by passing through columns of activated alumina and Q-5 copper catalyst. Mehylaluminoxane was obtained as a 30 weight percent solution in toluene from Albamarle and was stored in a refrigerator in argon glove box. NMR spectra were obtained on Varian Unity Plus 500-MHz FT-NMR spectrometer using solvent $CD_3SO_2CD_3$. Infrared spectra were measured using a Mattson Galaxy Series FTIR 5000 spectrophotometer running FIRST software. GPCs were done in TCB @135 °C on a Waters Associates 150C High Temperature Gel permeation chromatograph equipped with three Polymer Laboratory mixed-bed type D columns in 1,2,4-trichlorobenzene at 135 °C using triple detector (Precision Detector 15 °C fixed-angle light scattering; Viscotek high-temperature viscometer; Waters 150 °C high temperature DRI). DSC data were obtained on a TA Instrument 2920 calorimeter using a scan rate of 10 °C/min.

Preparation of 1,1′bis(1-hydrobenzimidazol-2yl)methane (HBBIM).

In a 1 L round bottom flask, 24.06 g (0.23 mol) of malonic acid, 50.00 g of (0.46 mol) of phenylenediamine, and 85 g of polyphosphoric acid were added sequentially. The mixture was heated at 210 °C using a Dean-Stark trap for 3.5 h with stirring. The mixture was cooled to 150 °C and then poured into 1 L of water. A blender was used to grind the slurry. The mixture was then neutralized to pH 8 with ammonium hydroxide. The solid was collected by filtration and repeatedly washed with water. The solid was dried to constant weight using vacuum at 80 °C for 24 h. $C_{15}H_{12}N_4$, FW = 248.29. Yield: 33.1 g, 58%. mp 331 °C (dec.); ^{1}H NMR ($CD_3SO_2CD_3$): δ = 7.53(q, J = 3.1 Hz, 4H), 7.16(q, J = 3.1 Hz, 4H), 4.54(s, 2H), 2.51(s, 2H); ^{13}C NMR ($CD_3SO_2CD_3$) δ 150.6, 139.1, 121.9, 115.1, 29.7. - R_f = 0.28 (Etoac), FD MS 248.2. IR (KBr pellet, cm^{-1}) 1622, 1590, 1544, 1530, 1487, 1435 s, 1309, 1271 s, 1031 s, 999, 738 s, 618, 466. Anal. Calcd for N 22.57, H 4.87, C 72.56. Found: N 22.57, H 4.86, C 72.36.

Preparation of 1,1′bis(1-butylbenzimidazol-2yl)pentane, (Bu₃BBIM) ligand.

A 23.00 g (0.093 mol) quantity of **4** was placed in a 300 mL round bottom flask with a side arm. This was followed by the addition of 25 mL of DMSO. The flask was fitted with a bubbler. Then under a flow of nitrogen 6.0 g of sodium hydride (60% dispersion in mineral oil) was added over one hour with stirring.

Then 25.00 mL (0.220 mol) of 1-iodobutane was added dropwise over one hour. The reaction mixture was left stirring under nitrogen for 48 hours. The reaction mixture was quenched with water and then an additional 400 mL of water was added. After stirring for one half hour a biphasic solution was obtained. The organic layer was extracted with cyclohexane and was washed with water. The volatiles were removed under reduced pressure to leave a dark oil. The oil was chromatographed on silica gel with methylene chloride as the eluent. The solvent was removed under reduced pressure to give a very pale-pink oil which crystallized upon standing. $C_{27}H_{36}N_4$, FW = 416.61. Yield: 20.44 g, 67%. mp 82-83 °C ^{1}H NMR (CDCl$_3$): δ = 7.79(m, 2H), 7.24(m, 6H), 4.88(t, J = 7.9 Hz, 1H), 4.16(d sp, J = 5.0 Hz, J = 40.9 Hz, 4H), 2.59(m, 2H), 1.44(m, 4H), 1.16(m, 4H), 1.10(m, 2H), 0.99(m, 2H), 0.89(t, J = 6.8 Hz, 3H), 0.61(t, J = 7.0 Hz, 6H). ^{13}C NMR (CDCl$_3$): δ = 151.8, 142.4, 135.6, 122.6, 122.0, 119.6, 109.7, 44.0, 40.9, 31.5, 31.2, 30.1, 22.5, 20.0, 14.0, 13.4. - R_f = 0.73 (Etoac) – FD MS 416.1, IR (KBr pellet, cm^{-1}) 3050 m, 2955 m, 2931 m, 2862 m, 1613 w, 1501 m, 1458 s, 1400 s, 1331 m, 1285 m, 1008 m, 933 m, 743 s, 434 w.

Preparation of the catalyst [1,1′bis(1-butylbenzimidazol-2yl)pentane] nickel(II) dichloride [Ni(Bu₃BBIM)Cl₂]

A 100-mg (0.42 mmol) quantity of NiCl$_2$·6H$_2$O was dissolved in 20 mL of absolute ethanol to give a yellow-green solution. Then 190 mg (0.46 mmol) of tributBBIM was added followed by the addition of 1 mL of triethylorthoformate. The solution was heated to gentle reflux for ca. 5 min. Upon cooling violet dichroic blades formed and 227 mg of solid was collected by filtration and washed with triethylformate followed by pentane. (98%), $C_{27}H_{36}Cl_2N_4Ni$, FW = 546.22; mp 324-325 °C (decomp.); X-ray crystallographic data: monoclinic, P2(1), a = 14.0690 Å, b = 14.1050 Å, c = 14.3130 Å, α = 90°, β = 97.220°, γ = 90°, V = 2817.80.

Preparation of 2,2′-bis[2-(1-hydrobenzimidazol-2yl)]biphenyl, (HBBIL).

In a 1 L round bottom flask, 35.00 g (0.145 mol) of diphenic acid, 31.25 g of (0.289 mol) phenylenediamine, and 50 g of polyphosphoric acid were added sequentially. The mixture was heated at 210 °C using a Dean-Stark trap for 3.5 h with stirring. The mixture was cooled to 150 °C and then poured into 1 L of water. A blender was used to grind the slurry. The mixture was then neutralized to pH 8 with ammonium hydroxide. The solid was collected by filtration and repeatedly washed with water. The solid was dried to constant weight using vacuum at 80 °C for 24 h. $C_{26}H_{18}N_4$, FW = 386.45. Yield: 53.98 g, 96%. mp 295 °C - ^{1}H NMR (CD$_3$SO$_2$CD$_3$): δ = 7.73 (d, J = 7.5 Hz, 2H), 7.50(q, J = 3.5 Hz, 4H), 7.34(t, J = 7.0 Hz, 2H), 7.22(t, J = 7.0 Hz, 2H), 7.16(q, J = 3.0 Hz, 4H), 6.97(d, J = 7.5 Hz). ^{13}C NMR (CD$_3$SO$_2$CD$_3$): δ = 153.2, 141.0, 138.7,

130.6, 130.4, 130.0, 128.9, 127.4, 122.1, 115.2. - R_f = 0.36 (Etoac) – FD MS 387.3. IR (KBr pellet, cm^{-1}) 1709 m, 1621 m, 1431 s, 1369 w, 1274 m, 744 s, 531 w, 429 w. Anal. Calcd for N 14.50, H 4.69, C 80.81. Found: N 14.41, H 4.79, C 80.65.

Preparation of 2,2´-bis[2-(1-ethylbenzimidazol-2yl)]biphenyl · water, (Et₂BBIL · water).

A 12.09 g (0.031 mol) quantity of 2,2´-bis[2-(1-hydrobenzimidazol-2yl)]biphenyl, (HBBIL). was placed in a 250 mL round bottom flask with a side arm. This was followed by the addition of 100 mL of DMSO. The flask was fitted with a bubbler. Then under a flow of nitrogen 6.0 g of sodium hydride (80% dispersion in mineral oil) was added over one hour with stirring. Then 5.20 mL (0.065 mol) of iodoethane was added drop wise over one hour. The reaction mixture was left stirring under nitrogen overnight. The reaction mixture was quenched with water and then an additional 400 mL of water was added. After stirring for one half hour a pale-pink solid separated from the solution. The solid was collected by filtration and repeatedly washed with water. The solid was dried in a vacuum oven for 24 hours at 60 °C. The solid was dissolved in 35 mL of methylene chloride to give a dark solution. Then 150 mL of pentane was added to give a cloudy solution. The mixture was placed in an ice bath for 30 min. After which a dark sticky precipitate formed. The cloudy solute was decanted and the solvent was removed under reduced pressure to give a white crystalline solid. $C_{30}H_{28}N_4O_1$, FW = 460.58. Yield: 13.12 g, 92%. mp 166 (soften) 172 °C (melt) - ^{1}H NMR (CDCl$_3$): δ = 7.60(d, J = 7.0 Hz, 2H), 7.41(d, J = 6.5 Hz, 2H), 7.30(m, 6H), 7.23(m, 6H), 3.67(brd s, 4H), 1.19(t, J = 7.0 Hz, 6H). - ^{13}C NMR (CDCl$_3$): δ = 152.5, 143.4, 141.1, 134.5, 131.5, 130.7, 130.0, 129.4, 127.1, 122.2, 121.8, 120.0, 110.0, 38.9, 14.8. - R_f = 0.62 (Etoac) – FD MS 442.0 (less H$_2$O), IR (KBr pellet, cm^{-1}) 3378 m (H$_2$O), 3062 m, 2968 m, 2931 m, 2871 m, 1643 s, 1447 s, 1329 s, 1276 s, 1128 m, 743 s, 476 w. Anal. Calcd for N 12.16, H 6.13, C 78.23. Found: N 11.91, H 6.14, C 79.16.

Preparation of ± 2,2'-bis[2-(1-ethylbenzimidazol-2yl)]biphenyl]nickel(II) dichloride [Ni(Et₂BBIL)Cl₂]

A 100-mg (0.42 mmol) quantity of NiCl$_2$·6H$_2$O was dissolved in a mixture consisting of 15 mL of ethanol and 1.5 mL of triethylorthoformate to give a yellow-green solution. After the addition of 60 mg (0.14 mmol) of 2,2'-bis[2-(1-ethylbenzimidazol-2yl)]biphenyl the mixture was warmed. Upon cooling, a bright-blue crystalline solid precipitated. The precipitate was collected by filtration and was then redissolved in 5 mL of warm nitromethane. The solution was filtered and upon standing yielded bright-blue prisms. FW = 572.16; X-ray crystallographic data: Space group = $P2_1P2_1P2_1$ a = 9.854(1) Å, b = 16.695(2) Å, c = 16.844(2) Å, V = 2771.05(5) Å^3, Z = 4, R = 0.0469.

Polymer Syntheses

Polymerization of 2-Norbornene Using Ni(II) Bis-benzimidazole Complexes

A slurry of Ni- bis-benzimidazole complex was prepared by suspending Ni(tributylBBIM)Cl$_2$ (FW 546.22) (8.0 mg, 1.46×10^{-2} mmol) in 6 g toluene, followed by activation with the appropriate amount of 30% MAO to obtain desired Al/Ni ratio (Table 1 and 2). The product is generally a dark suspension. A 1g quantity of norbornene was added and the solution was stirred (stirring rate 500 RPM) at 25°C for the appropriate reaction time. The reaction was quenched by pouring the mixture into 200-mL methanol/10% HCl (50:1). The colorless product was isolated by filtration, washed carefully with methanol, and dried in vacuum. The yield was determined and the product was characterized by ^{13}C NMR, IR and GPC.

Summary

Late transition metal catalyst complexes are of increasing interest with respect to oligomerization and polymerization of olefins. Previously, we have reported the synthesis of copper (II) bis-benzimidazole complexes. These complexes, upon activation with MAO, polymerize ethylene to highly linear polyethylene. Analogous Nickel (II) complexes have also been prepared and found to polymerize the cyclic olefin norbornene. The polynorbornene synthesized was characterized by GPC, ^{13}C NMR and IR. GPC molecular weights of the polynorbornene polymers were determined using the LS/viscometry/DRI triple detector system and were found to be very high (Mn 148,000 to 288,000 and Mw 587,000 to 797,000). The ^{13}C NMR suggest that the polynorbornene is a vinyl addition product (*i. e.,* 2,3-linked); no ring-opened product was observed.

Acknowledgement

We would like to thank B. Liang for NMR, M. Varma-Nair for TGA and J. A. Olkusz for GPC measurements and useful discussion. We would also like to thank Tom Coolbaugh for useful comments.

Literature Cited

1.   Ittel, S. D.; Johnson, L. K.; Brookhart, M. *Chem. Rev.* **2000**, 100, 1169-1203.

2. Britovsek, G. J. P.; Gibson, V. C.; Wass, D. F. *Angew. Chem. Int. Ed.* **1999**, 38, 428-447.

3. (a) Stibrany, R. T.; Patil, A. O.; Zushma. *Polym. Mater. Sci. Eng.* **2002**, 86, 323. (b) Stibrany, R. T.; Schulz, D. N.; Kacker, S.; Patil, A. O.; Baugh, L. S.; Rucker, S. P.; Zushma, S.; Berluche, E.; Sissano, J. A. *Polym. Mater. Sci. Eng.* **2002**, 86, 325. (c) Stibrany, R. T.; Schulz, D. N.; Kacker, S.; Patil, A. O. U. S. Patent 6,037,297, 2000 (March 14, 2000); U. S. Patent 6,417,303, 2002 (July 9, 2002) (d) Stibrany, R. T., U. S. Patent 6,180,788, 2001 (e) Stibrany, R. T.; Schulz, D. N.; Kacker, S.; Patil, A. O., PCT Int. Appli. WO 99/30822 (Exxon Research & Engineering Company), June 24, 1999.

4. U. S. Patent 2,721,189 (1954) to Du Pont.

5. Ivin, K. J.; Mol, J. C. *"Olefin Metathesis and Metathesis Polymerization"*, Academic Press, San Diego, CA 1997, pp. 407-410.

6. Kennedy, J. P.; Makowsky, H. S. *J. Macromol. Sci., Chem.* **1967**, A1, 345.

7. Gaylord, N. G.; Mandal, B. M.; Martan, M. *J. Poly. Sci., Polym. Lett. Ed.* **1976**, 14, 555.

8. Gaylord, N. G.; Deshpande, A. B.; Mandal, B. M.; Martan, M. *J. Macromol. Sci., Chem.* **1977**, A11, 1053.

9. Gaylord, N. G.; Deshpande, A. B. *J. Poly. Sci., Polym. Lett. Ed.* **1976**, 14, 613.

10. Hauthal, H. G. *Nachr. Chem. Tech. Lab.* **1995**, 43, 822.

11. Vennen, H. *Future - Hoechst Magazin* IV/**1995**, 52.

12. Mast, C.; Krieger, M.; Dehnicke, K.; Greiner, A. *Macromol. Rapid Commun.* **1999**, 20(4), 232. *Polym. Mater. Sci. Eng.* **1999**, 80, 423.

13. Goodall, B. L.; Barnes, D. A.; Benedikt, G. M.; McIntosh, L. H.; Rhodes, L. F. *Polym. Mater. Sci. Eng.* **1997**, 76, 56.

14. Wu, Q.; Lu, Y.; Lu, Z.. *Polym. Mater. Sci. Eng.* **1999**, 80, 483.

15. Kaminsky, W.; Bak, A.; Steiger, R. *J. Mol. Cata.* **1992**, 74, 109.

16. Kaminsky, W. *Macromol. Chem. Phys.* **1996**, 197, 3907.

17. Heinz, B. S.; Alt, F. P.; Heitz, W. *Macromol. Rapid Commun.* **1998**, 19(5), 251.

18. Abu-S., Adnan S.; Rieger, B. *J. Mol. Catal. A: Chem.* **1998**, 128(1-3), 239-243.

19. Heinz, B. S.; Alt, F. P.; Heitz, W. *Macromol. Rapid Commun.* **1998**, 19(5), 251.

20. Vyas, P., C.; Oza, C., K.; Goyal, A., K. *Chem. and Ind.* **1980,** 287.

Catalysts Based on Copper Complexes

Copper-Based Olefin Polymerization Catalysts

Robert T. Stibrany, Abhimanyu O. Patil, and Stephen Zushma

Corporate Strategic Research, ExxonMobil Research and Engineering Company, 1545 Route 22 East, Annandale, NJ 08801

Novel copper (II) bis-benzimidazole complexes were prepared. These complexes were synthesized using three steps: condensation, alkylation and complexation. The complexes were characterized by melting point, elemental analysis, IR, ^{1}H and ^{13}C NMR spectroscopy and X-ray crystallography. These complexes, upon activation with MAO, polymerize ethylene. The polyethylene obtained was typically a high molecular weight and a high melting polymer. The ^{13}C NMR of polyethylene prepared by Cu BBIM/MAO contains no detectable branches. The ethylene homopolymerization data are consistent with a single-site coordination/insertion mechanism based upon the high degree of linearity and narrow MWD in the product.

There has been considerable recent activity in the area of late transition metal polymerization catalysts (*1-5*). In the last two decades, metallocenes have revolutionized the commercial polymerization of polyolefins. In many cases, these catalysts are now used in place of Ziegler-Natta catalysts to produce better performing high-density polyethylene (HDPE) and linear low-density polyethylene (LLDPE). Over the past few years, a trend in catalyst development has moved from modification of Group 4-metallocene catalysts to a search for new generation catalysts, such as non-metallocene catalysts and late transition-metal catalysts (*1-5*). The polymerization activity with the non-metallocene catalysts has thus far been generally lower than that with the metallocene catalysts. However, some non-metallocene catalysts [i.e. zirconium complexes involving a bis(salicylaldiimine) ligand (*6*) and iron(II) complexes based on the tridentate 2,6-bis(imino)pyridine ligand] show high activity for ethylene polymerization (*7-9*). Great advances have been made in recent years in the field of mid- and late-transition metal catalysts for polymerization. The catalysts reported in the literature to this point are based on Group 8-10 metals (Fe, Co, Ni, and Pd) (*1-5*). Previously, no olefin polymerization catalysts based on the Group 11 metals had been prepared. This paper details the synthesis of copper (II) bis-benzimidazole complexes and Cu catalyzed homopolymerizations of ethylene (*10-12*).

The copper complexes are prepared in three synthetic steps as illustrated in Scheme 1. The first step is the condensation reaction of a dicarboxylic acid in phosphoric acid with two equivalents of phenylenediamine (*13*). Condensation of malonic acid yields the methylene bridged bisbenzimidazole, **4**. Condensation of diphenic acid yields the 2,2' biphenyl bridged bis-benzimidazole, **5**. The next synthetic step is alkylation of the bisbenzimidazole. Compound **4** can be alkylated sequentially up to four times, once each at the protonated nitrogens and twice at the methylene bridgehead carbon. In this study **4** was alkylated three times with ethyl iodide and butyl iodide to prepare alkylated products **6** and **7**. Compound **5** was alkylated twice, once at each protonated nitrogen. The alkylation was carried out with ethyl iodide and octyl iodide to prepare alkylated products **8** and **9**.

Finally, the alkylated bisbenzimidazoles are metalated using $CuCl_2 \cdot 2H_2O$ in a mixture of ethanol and triethylorthoformate. Triethylorthoformate is added to effect the dehydration of the reaction mixture and aid in the crystallization of the metal complex. The solid crystalline complexes are then isolated by filtration. The alkylated bisbenzimidazole **6** and **7** were reacted with $CuCl_2 \cdot 2H_2O$ to prepared complexes **10** and **11** respectively. The alkylated bisbenzimidazole **7** were reacted with $CuBr_2$ to prepared complex **12**. The alkylated bisbenzimidazole **8** and **9** were reacted with $CuCl_2 \cdot 2H_2O$ to prepared complexes **13** and **14** respectively. Figure 1 shows an X-ray (ORTEP) structure for **13** (*14*).

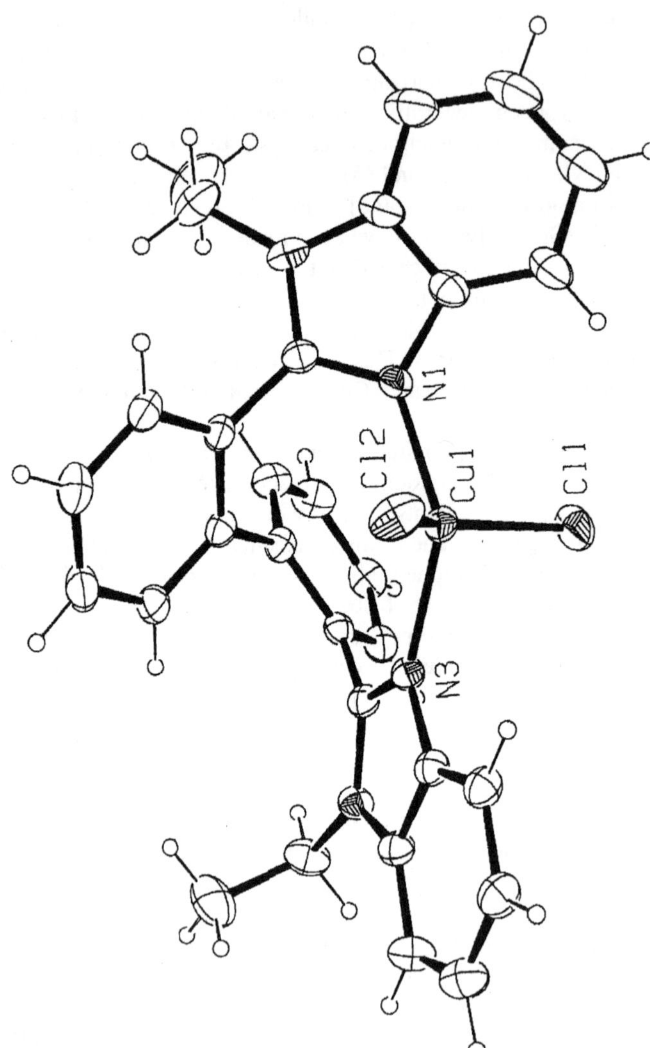

*Figure 1. (ORTEP) of **13** showing the atom numbering scheme. Thermal ellipsoids are drawn at the 30% probability level. 1,2-DCB has been omitted for clarity.*

Condensation

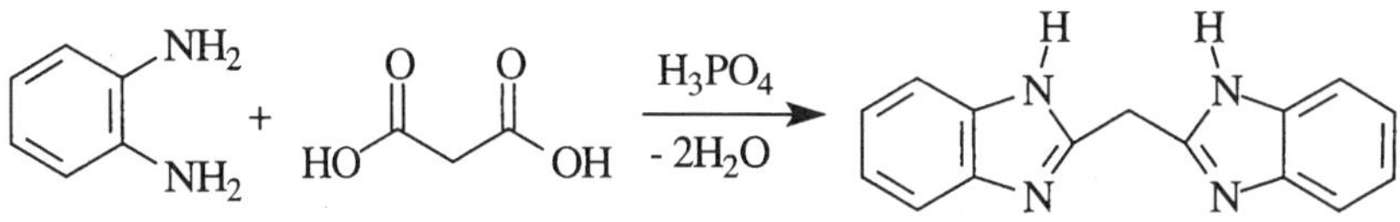

$$\text{H}_3\text{PO}_4,\ -2\text{H}_2\text{O}$$

1 + **2** → **4**

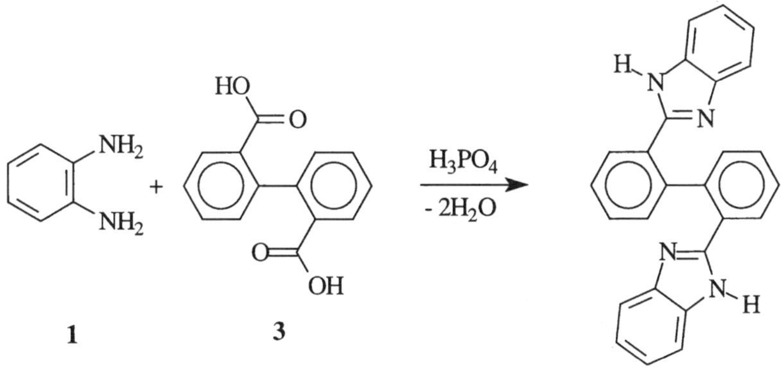

$$\text{H}_3\text{PO}_4,\ -2\text{H}_2\text{O}$$

1 + **3** → **5**

Alkylation

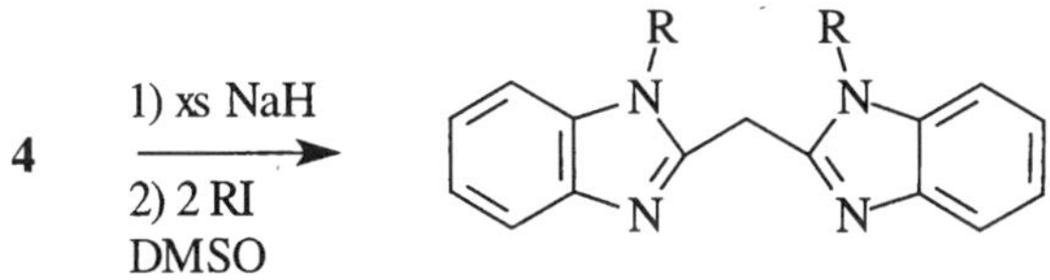

4 1) xs NaH 2) 2 RI DMSO →

6: R = C_2H_5
7: R = n-C_4H_9

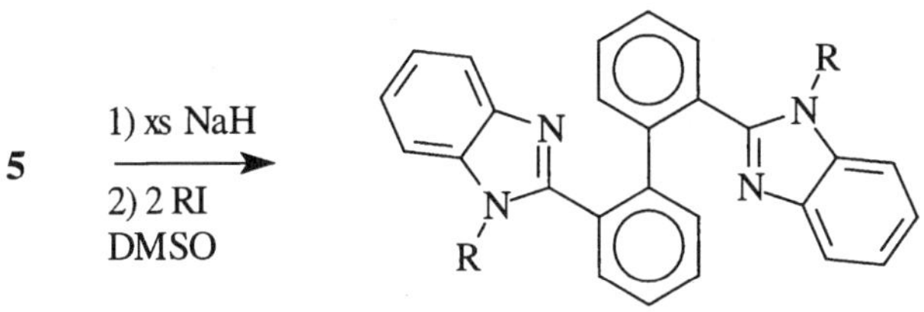

5 1) xs NaH 2) 2 RI DMSO →

8: R = C_2H_5
9: R = C_8H_{17}

Metalation

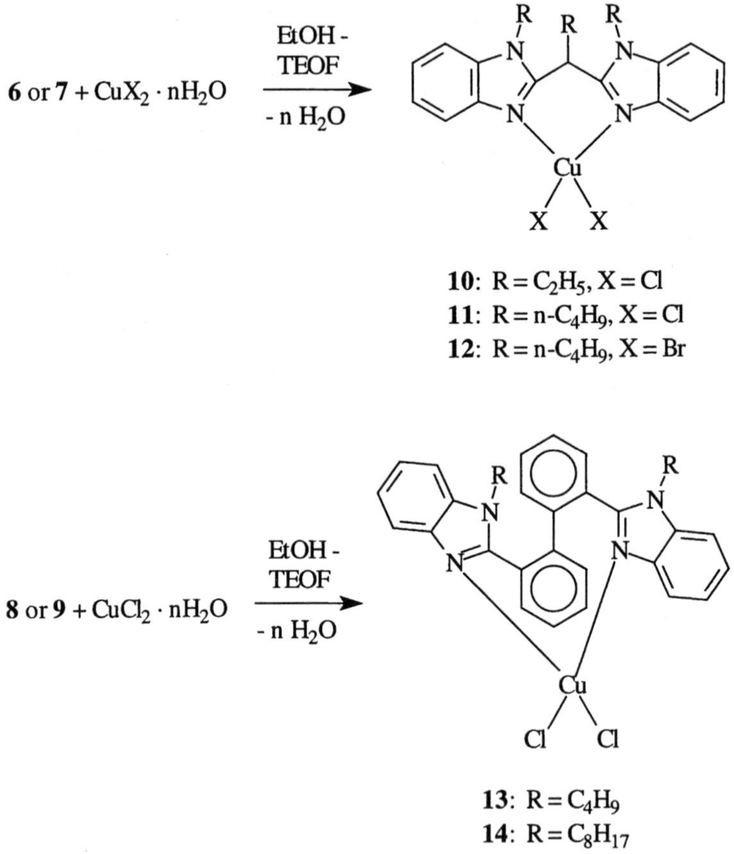

6 or 7 + CuX$_2$ · nH$_2$O $\xrightarrow[\text{- n H}_2\text{O}]{\text{EtOH - TEOF}}$

10: R = C$_2$H$_5$, X = Cl
11: R = n-C$_4$H$_9$, X = Cl
12: R = n-C$_4$H$_9$, X = Br

8 or 9 + CuCl$_2$ · nH$_2$O $\xrightarrow[\text{- n H}_2\text{O}]{\text{EtOH - TEOF}}$

13: R = C$_4$H$_9$
14: R = C$_8$H$_{17}$

Scheme 1. Synthetic steps for the preparation of Cu(II) complexes.

Experimental Section

General Conditions. Alkylations utilizing NaH were conducted under nitrogen using Schlenk techniques. Dimethyl sulfoxide was dried over CaH_2. All other reagents were used as received or purified by standard methods. NMR spectra were recorded using a Bruker AVANCE 400 Ultrashield spectrometer. Melting points were determined with a hot-stage apparatus and are uncorrected. Infrared spectra were measured using a Mattson Galaxy Series 5000 spectrophotometer. Abbreviations used are as follows: TEOF for triethylorthoformate, Etoac for ethylacetate, ACN for acetonitrile, THF for tetrahydrofuran, 1,2DCB for 1,2-dichlorobenzene, qt for quintet, sp for septet. Elemental analyses were carried out by Galbraith Laboratories, Knoxville, TN.

Bis(1-Hydrobenzimidazoles).

Synthesis of compounds 4 and 5.

1,1′bis(1-hydrobenzimidazol-2yl)methane, HBBIM (4). In a 1 L round bottom flask, 24.06 g (0.23 mol) of **2**, 50.00 g of (0.46 mol) **1**, and 85 g of polyphosphoric acid were added sequentially. The mixture was heated at 210 °C using a Dean-Stark trap for 3.5 h with stirring. The mixture was cooled to 150 °C and then poured into 1 L of water. A blender was used to grind the slurry. The mixture was then neutralized to pH 8 with ammonium hydroxide. The solid was collected by filtration and repeatedly washed with water. The solid was dried to constant weight using vacuum at 80 °C for 24 h. $C_{15}H_{12}N_4$, FW = 248.29. Yield: 33.1 g, 58%. mp 331 °C (dec.); ^1H NMR ($CD_3SO_2CD_3$): δ = 7.53(q, J = 3.1 Hz, 4H), 7.16(q, J = 3.1 Hz, 4H), 4.54(s, 2H), 2.51(s, 2H); ^{13}C NMR ($CD_3SO_2CD_3$) δ 150.6, 139.1, 121.9, 115.1, 29.7. - R_f = 0.28 (Etoac), FD MS 248.2. IR (KBr pellet, cm^{-1}) 1622, 1590, 1544, 1530, 1487, 1435 s, 1309, 1271 s, 1031 s, 999, 738 s, 618, 466. Anal. Calcd for N 22.57, H 4.87, C 72.56. Found: N 22.57, H 4.86, C 72.36.

2,2′-bis[2-(1-hydrobenzimidazol-2yl)]biphenyl, HBBIL (5). In a 1 L round bottom flask, 35.00 g (0.145 mol) of **3**, 31.25 g of (0.289 mol) **1**, and 50 g of polyphosphoric acid were added sequentially. The mixture was heated at 210 °C using a Dean-Stark trap for 3.5 h with stirring. The mixture was cooled to 150 °C and then poured into 1 L of water. A blender was used to grind the slurry. The mixture was then neutralized to pH 8 with ammonium hydroxide. The solid was collected by filtration and repeatedly washed with water. The solid was dried to constant weight using vacuum at 80 °C for 24 h. $C_{26}H_{18}N_4$, FW = 386.45. Yield: 53.98 g, 96%. mp 295 °C - ^1H NMR ($CD_3SO_2CD_3$): δ = 7.73 (d, J = 7.5 Hz, 2H), 7.50(q, J = 3.5 Hz, 4H), 7.34(t, J = 7.0 Hz, 2H), 7.22(t, J = 7.0 Hz, 2H), 7.16(q, J = 3.0 Hz, 4H), 6.97(d, J = 7.5 Hz). ^{13}C NMR ($CD_3SO_2CD_3$): δ = 153.2, 141.0, 138.7, 130.6, 130.4, 130.0, 128.9, 127.4, 122.1, 115.2. - R_f =

0.36 (Etoac) – FD MS 387.3. IR (KBr pellet, cm^{-1}) 1709 m, 1621 m, 1431 s, 1369 w, 1274 m, 744 s, 531 w, 429 w. Anal. Calcd for N 14.50, H 4.69, C 80.81. Found: N 14.41, H 4.79, C 80.65.

Alkylation of bis(1-hydrobenzimidazoles).

Synthesis of Compounds 6-9.

1,1′bis(1-ethylbenzimidazol-2yl)propane, Et$_3$BBIM (6). A 19.10 g (0.077 mol) quantity of **4** was placed in a 500 mL round bottom flask with a side arm. This was followed by the addition of 30 mL of DMSO. The flask was fitted with a bubbler. Then under a flow of nitrogen 7.0 g of sodium hydride (80% dispersion in mineral oil) was added over one hour with stirring. Then 18.50 mL (0.231 mol) of iodoethane was added dropwise over one hour. The reaction mixture was left stirring under nitrogen overnight. The reaction mixture was quenched with water and then an additional 150 mL of water was added. After stirring for one half hour a brown solid seperated from the solution. The solid was collected by filtration and repeatedly washed with water. The solid was dried in a vacuum oven for 24 hours at 60 °C. The solid was dissolved in a sufficient amount of acetone to give a yellow-brown solution. Then enough cyclohexane was added to give a cloudy solution. The mixture was refluxed on a steam bath for 10 min. followed by immersion in an ice bath for 15 min. After which a dark sticky precipitate formed. The precipitate was removed by filtration. The filtrate was concentrated to dryness under reduced pressure to give a very pale-pink solid. C$_{21}$H$_{24}$N$_4$, FW = 332.45. Yield: 18.91 g, 74%. mp 178 °C (dec.); ^{1}H NMR (CDCl$_3$): δ = 7.80(m, 2H), 7.26(m, 6H), 4.78(t, J = 7.9 Hz, 1H), 4.32(q, J = 7.3 Hz, 4H), 2.63(qt, J = 7.5 Hz, 2H), 1.12(t, J = 7.3 Hz, 3H), 0.96(t, J = 7.2 Hz, 6H). ^{13}C NMR (CDCl$_3$): δ = 151.5, 142.5, 135.2, 122.7, 122.1, 119.6, 109.6, 42.4, 38.8, 24.9, 14.6, 12.7. R_f = 0.50 (Etoac), FD MS 332.2. IR (KBr pellet, cm^{-1}) 3054 w, 2969 w, 2935 w, 2871 w, 1612 w, 1460 s, 1405 s, 1372 m, 1330 m, 1259 m, 1127 m, 964 m, 750 s, 421 w. Anal. Calcd for N 16.85, H 7.28, C 75.87. Found: N 16.21, H 7.36, C 75.77.

1,1′bis(1-butylbenzimidazol-2yl)pentane, Bu$_3$BBIM (7). A 23.00 g (0.093 mol) quantity of **4** was placed in a 300 mL round bottom flask with a side arm. This was followed by the addition of 25 mL of DMSO. The flask was fitted with a bubbler. Then under a flow of nitrogen 6.0 g of sodium hydride (60% dispersion in mineral oil) was added over one hour with stirring. Then 25.00 mL (0.220 mol) of 1-iodobutane was added dropwise over one hour. The reaction mixture was left stirring under nitrogen for 48 hours. The reaction mixture was quenched with water and then an additional 400 mL of water was added. After stirring for one half hour a biphasic solution was obtained. The organic layer was extracted with cyclohexane and was washed with water. The volatiles were removed under reduced pressure to leave a dark oil. The oil was chromatographed on silica gel with methylene chloride as the eluent. The solvent was removed under reduced pressure to give a very pale-pink oil which crystallized upon standing. $C_{27}H_{36}N_4$, FW = 416.61. Yield: 20.44 g, 67%. mp 82-83 °C - ^{1}H NMR (CDCl$_3$): δ = 7.79(m, 2H), 7.24(m, 6H), 4.88(t, J = 7.9 Hz, 1H), 4.16(d sp, J = 5.0 Hz, J = 40.9 Hz, 4H), 2.59(m, 2H), 1.44(m, 4H), 1.16(m, 4H), 1.10(m, 2H), 0.99(m, 2H), 0.89(t, J = 6.8 Hz, 3H), 0.61(t, J = 7.0 Hz, 6H). ^{13}C NMR (CDCl$_3$): δ = 151.8, 142.4, 135.6, 122.6, 122.0, 119.6, 109.7, 44.0, 40.9, 31.5, 31.2, 30.1, 22.5, 20.0, 14.0, 13.4. - R_f = 0.73 (Etoac) – FD MS 416.1, IR (KBr pellet, cm^{-1}) 3050 m, 2955 m, 2931 m, 2862 m, 1613 w, 1501 m, 1458 s, 1400 s, 1331 m, 1285 m, 1008 m, 933 m, 743 s, 434 w.

2,2′-bis[2-(1-ethylbenzimidazol-2yl)]biphenyl · water, Et$_2$BBIL · water (8). A 12.09 g (0.031 mol) quantity of **5** was placed in a 250 mL round bottom flask with a side arm. This was followed by the addition of 100 mL of DMSO. The flask was fitted with a bubbler. Then under a flow of nitrogen 6.0 g of sodium hydride (80% dispersion in mineral oil) was added over one hour with stirring. Then 5.20 mL (0.065 mol) of iodoethane was added drop wise over one hour. The reaction mixture was left stirring under nitrogen overnight. The reaction mixture was quenched with water and then an additional 400 mL of water was added. After stirring for one half hour a pale-pink solid separated from the solution. The solid was collected by filtration and repeatedly washed with water. The solid was dried in a vacuum oven for 24 hours at 60 °C. The solid was dissolved in 35 mL of methylene chloride to give a dark solution. Then 150 mL of pentane was added to give a cloudy solution. The mixture was placed in an ice bath for 30 min. After which a dark sticky precipitate formed. The cloudy solute was decanted and the solvent was removed under reduced pressure to give a white crystalline solid. $C_{30}H_{28}N_4O_1$, FW = 460.58. Yield: 13.12 g, 92%. mp 166 (soften) 172 °C (melt) - ^{1}H NMR (CDCl$_3$): δ = 7.60(d, J = 7.0 Hz, 2H), 7.41(d, J = 6.5 Hz, 2H), 7.30(m, 6H), 7.23(m, 6H), 3.67(brd s, 4H), 1.19(t, J = 7.0 Hz, 6H). - ^{13}C NMR (CDCl$_3$): δ = 152.5, 143.4, 141.1, 134.5, 131.5, 130.7, 130.0, 129.4, 127.1, 122.2, 121.8, 120.0, 110.0, 38.9, 14.8. - R_f = 0.62 (Etoac) – FD MS 442.0 (less H$_2$O), IR (KBr pellet, cm^{-1}) 3378 m (H$_2$O), 3062 m, 2968 m, 2931 m, 2871 m, 1643 s, 1447 s, 1329 s, 1276 s, 1128 m, 743 s, 476 w. Anal. Calcd for N 12.16, H 6.13, C 78.23. Found: N 11.91, H 6.14, C 79.16.

2,2′-bis[2-(1-octylbenzimidazol-2yl)]biphenyl, Oct$_2$BBIL (9). In an argon glovebox, a 16.19 g (0.042 mol) quantity of **5** was placed in a 300 mL round bottom flask. A suspension was obtained after the addition of 60 mL of THF. Then 3.2 g of sodium hydride powder was added over a half-hour period with stirring. After cooling 14.75 mL (0.082 mol) of 1-iodooctane was added dropwise over a half-hour period. After an hour a gray-white precipitate began to form. The mixture was allowed to stir an additional 12 hours. The reaction mixture was removed from the glove box and slowly quenched with water under nitrogen. The reaction mixture was extracted from water with methylene chloride. The methylene chloride was then dried over sodium sulfate and filtered. The methylene chloride was removed under high vacuum to give a pale-brown oil. $C_{42}H_{50}N_4$, FW = 610.88. Yield: 24.03 g, 95%. oil - ^{1}H NMR (CDCl$_3$): δ = 7.58(m, 2H), 7.41(m, 2H), 7.31(m, 4H), 7.23(m, 8H), 3.54(br s, 4H), 1.64(br s, 4H), 1.19(br m, 20 H), 0.86(t, J = 6.9 Hz, 6H). - ^{13}C NMR (CDCl$_3$): δ = 151.6, 142.3, 140.0, 133.7, 130.4, 129.7, 128.5, 128.4, 126.0, 121.1, 120.7, 118.9, 109.0, 43.1, 30.7, 28.4, 28.1, 27.9, 25.6, 21.6, 13.1. - R_f 0.77 = (Etoac), IR (film on KBr plates, cm^{-1}) 3060 w, 2954 m. 2928 s, 2855 m, 1613 w, 1453 s, 1392 m, 1282 m, 1159 m, 744 s. Anal. Calcd for N 9.17, H 8.25, C 82.58. Found: N 9.07, H 8.25, C 82.11.

Metalation of bis(alkylbenzimidazoles).

Synthesis of Compounds 10-14.

[1,1′bis(1-ethylbenzimidazol-2yl)propane]copper(II) dichloride, Cu(Et$_3$BBIM)Cl$_2$ (10).
In an argon glovebox, a 8.11 g (0.0476 mol) quantity of copper (II) dichloride dihydrate was dissolved in 500 mL of absolute ethanol. This was followed by the addition of 100 mL of TEOF. Then 14.37 g (0.0432 mol) of **6** was added to the solution. The solution was heated at reflux for a half-hour. Upon cooling the bright yellow precipitate was collected by filtration and washed with TEOF followed by pentane. The solid was dried under high vacuum. $C_{21}H_{24}Cl_2CuN_4$, FW = 466.90. Yield: 19.20 g, 95%. - mp (stage) 262 °C (sharp, dec.), (DSC) 266 °C; IR (KBr pellet, cm^{-1}) 3053 w, 2976 m, 2935 w, 2874 w, 1615 w, 1511 s, 1459 s, 1332 m, 1271 m, 1080 m, 974 m, 753 s, 560 w, 426 w. Anal. Calcd for N 12.00, H 5.18, C 54.02, Cu 13.61, Cl 15.19. Found: N 11.88, H 5.34, C 53.73, Cu 13.19, Cl 14.92.

[1,1′bis(1-butylbenzimidazol-2yl)pentane]copper(II) dichloride, Cu(Bu$_3$BBIM)Cl$_2$ (11).
In an argon glovebox, a 5.41 g (0.0317 mol) quantity of copper (II) dichloride dihydrate was dissolved in 300 mL of absolute ethanol. This was followed by the addition of 60 mL of TEOF. Then 12.02 g (0.0289 mol) of **7** was added to the solution. The solution was heated at reflux for a half-hour. Upon cooling the bright yellow precipitate was collected by filtration and washed with TEOF followed by pentane. The solid was dried under high vacuum. $C_{27}H_{36}Cl_2CuN_4$, FW = 551.06. Yield: 15.20 g, 96%. mp (stage) 257 °C (sharp, decomp.). IR (KBr pellet, cm^{-1}) 3102 w, 3060 w, 3027 w, 2959 s, 2930 s, 2871 m, 2859 m, 1616 w, 1513 s, 1460 s, 745 s, 502 w. Anal. Calcd for N 10.17, H 6.59, C 58.85, Cu 11.53, Cl 12.87. Found: N 10.16, H 6.72, C 58.79, Cu 11.43, Cl 12.92.

***pro-R* [1,1′bis(1-butylbenzimidazol-2yl)pentane]copper(II) dibromide, Cu(Bu$_3$BBIM)Br$_2$ (12).**

In an argon glovebox, a 10.62 g (0.0475 mol) quantity of copper (II) dibromide was dissolved in 900 mL of absolute ethanol. This was followed by the addition of 180 mL of TEOF. Then 18.00 g (0.0432 mol) of **7** was added to the solution. The solution was heated at reflux for a half-hour. Upon cooling the maroon precipitate was collected by filtration and washed with TEOF followed by pentane. The solid was dried under high vacuum. C$_{27}$H$_{36}$Br$_2$CuN$_4$, FW = 639.96. Yield: 17.47 g, 63%. mp (stage) 221 °C. (KBr pellet, cm^{-1}) 3097 w, 3061 w, 3030 w, 2957 s, 2930 s, 2872 m, 2858 m, 1613 w, 1510 s, 1454 s, 755 s, 505 w. Anal. Calcd for N 8.75, H 5.67, C 50.67, Cu 9.94, Br 24.97. Found: N 8.65, H 5.59, C 50.11, Cu 10.00, Br 25.60.

***rac*-[2,2′-bis[2-(1-ethylbenzimidazol-2yl)]biphenyl]copper(II) dichloride · 1,2-dichlorobenzene], Cu(Et$_2$BBIL)Cl$_2$ (13) ·1,2DCB.**

In an argon glovebox, a 5.30 g (0.0311 mol) quantity of copper (II) dichloride dihydrate was dissolved in 200 mL of absolute ethanol. This was followed by the addition of 125 mL of TEOF. Then 12.50 g (0.0280 mol) of 8 was added to the solution. The solution was heated at reflux for a half-hour. Upon cooling the bright yellow precipitate was collected by filtration and washed with TEOF followed by pentane. The solid was dried under high vacuum to give 11.54 g of yellow-orange solid. The solid was redissolved in a mixture of methylene chloride and 1,2-dichlorobenzene. This solution was placed in a dessicator containing pentane to allow vapor diffusion. After several days yellow-green prisms precipitate and are collected by filtration and washed with pentane and finally dried under ambient argon. C$_{36}$H$_{30}$Cl$_4$CuN$_4$, FW = 724.02 - mp (stage) 150 °C (opaque, loss of 1,2DCB) 300-301 °C (melt) IR (KBr pellet, cm^{-1}) 3056 w, 2989 w, 1612 w, 1470 s, 1419 s, 1345 m, 1273 m, 1135 m, 960 m, 752 s, 553 m, 424 w. Anal. (dried under high vacuum) Calcd for N 9.71, H 4.54, C 62.45, Cu 11.01, Cl 12.29. Found: N 9.44, H 4.66, C 61.65, Cu 10.72, Cl 12.18.

***rac*-[[S-2,2′-bis[2-(1-octylbenzimidazol-2yl)]biphenyl]copper(II) dichloride [R-2,2′-bis[2-(1-octylbenzimidazol-2yl)]biphenyl]copper(II) dichloride], Cu(Oct$_2$BBIL)Cl$_2$ (14).**

In an argon glovebox, a 6.50 g (0.0382 mol) quantity of copper (II) dichloride dihydrate was dissolved in 100 mL of absolute ethanol. This was followed by the addition of 100 mL of TEOF. Then 21.19 g (0.0347 mol) of 9 was added to the solution. The solution was heated at reflux for a half-hour. Upon cooling the shimmering yellow precipitate was collected by filtration and washed with TEOF followed by pentane. The solid was dried under high vacuum. C$_{42}$H$_{50}$Cl$_2$CuN$_4$, FW = 745.33. Yield: 13.02 g, 50%. mp (stage) 148 °C (soften) 152 °C (melt), (DSC) 156 °C (1st melt) 172 °C (recryst) 193 °C (2nd melt); IR (KBr pellet, cm^{-1}) 3060 w, 2927 s, 2854 s, 1613 w, 1467 s, 1415 s, 1293 m,

1159 m, 1010 w, 746 s. Anal. Calcd for N 7.52, H 6.76, C 67.68, Cu 8.53, Cl 9.51. Found: N 7.46, H 6.76, C 67.43, Cu 8.46, Cl 9.42. d = 1.07(1)g cm^{-3} (floatation in a mixture of cyclohexane/$CF_2ClCFCl_2$).

Ethylene Polymerization using Cu(II) Bis-benzimidazole Catalysts

$$H_2C{=}CH_2 \xrightarrow[\text{MAO}]{\text{Cu(BBIM)}X_2} \left(\!\!\!\bigwedge\!\!\!\right)_n$$

Materials and Instruments: Methylaluminoxane was obtained as a 30 weight percent solution in toluene from Albemarle and was stored in a refrigerator in argon glove-box. Ethylene was obtained from BOC. Toluene was distilled from Na benzophenone or was dried by passing through alumina and Q-5 copper catalyst. Methanol, and THF were obtained from Aldrich. ^{1}H and ^{13}C NMR were recorded on a Varian Unity Plus 500 MHz FT-NMR spectrometer. The solvent used was d^4-o-dichlorobenzene. $Cr(acac)_3$ was used as a relaxation agent. Molecular weights were measured on a Waters Associates Gel Permeation Chromatography in trichloroethane using polyethylene standard. Thermal characterization was done using a Seiko SII DSC instrument with a heating rate of 10 °C/min.

Ethylene Polymerization:

A glass lined Parr reactor was loaded in an Ar glove box with 27.1 mg (0.047 mmol) of $Cu(Et_2BBIL)Cl_2$ (Catalysts **13**) followed by 30 mL of toluene to give a pale yellow partially dissolved solution. Next 2.0 mL of 30% MAO was added to give a nearly colorless solution (Al/Cu = 220). The Parr was sealed and taken to a hood containing the controller for the Parr and pressurized with 700 psig ethylene and polymerized at 80 °C for ~ 24 hours. The reaction was cooled, vented and quenched with MeOH. The product was collected by filtration, washed with MeOH and dried at 70 °C for 2 hours. Yiled: 16.1 g.

The polyethylene product was characterized using GPC, IR, NMR and DSC. The Mw of the polymer was 150,000 with MWD of 2.5. The melting point of the polyethylene was 138 °C. ^{13}C NMR spectra of the polymers show only one methylene peak at 29.9 ppm and ^{1}H NMR spectra of the polymers show only one methylene peak at 1.34 ppm. This suggests that polyethylene have perfectly linear structure (no branching). A ^{13}C NMR and ^{1}H NMR spectrum of the sample are shown in Figures 2 and 3. The IR spectrum of the sample showed characteristic methylene peak at 719 and 729 cm^{-1} (Figure 4). The above data clearly suggest that polyethylene microstructure is highly linear.

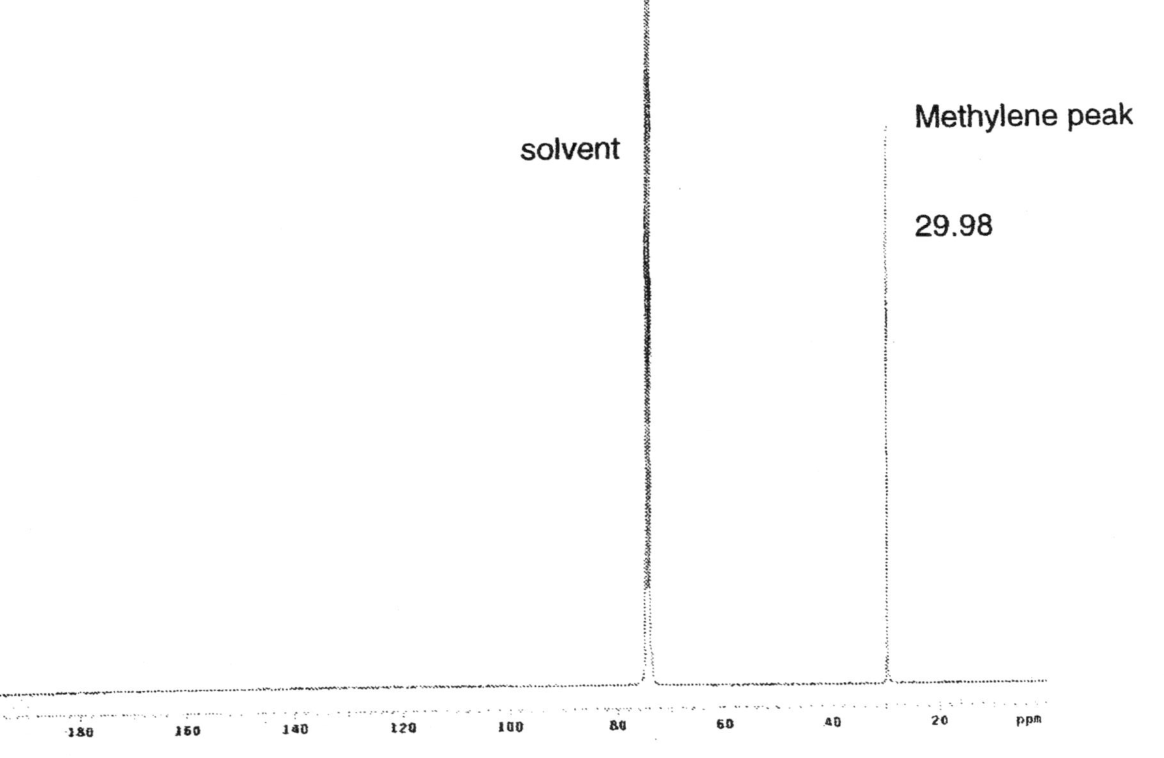

Figure 2. ^{13}C NMR spectrum of polyethylene sample.

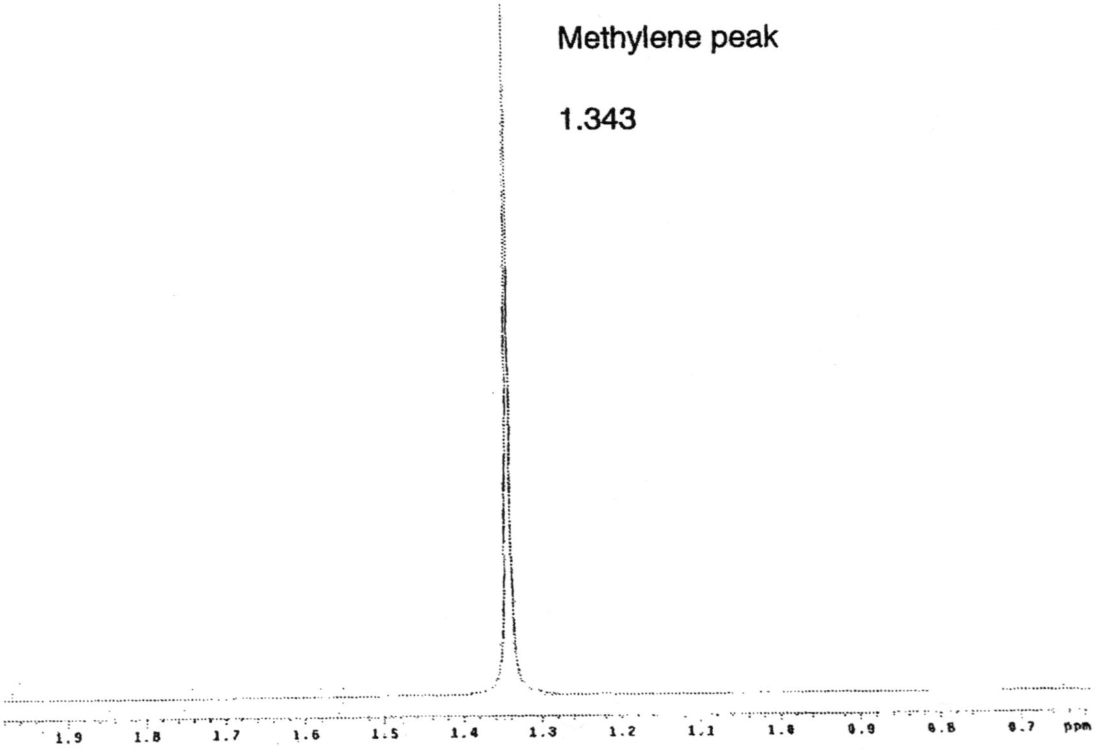

Figure 3. ^1H NMR spectrum of polyethylene sample.

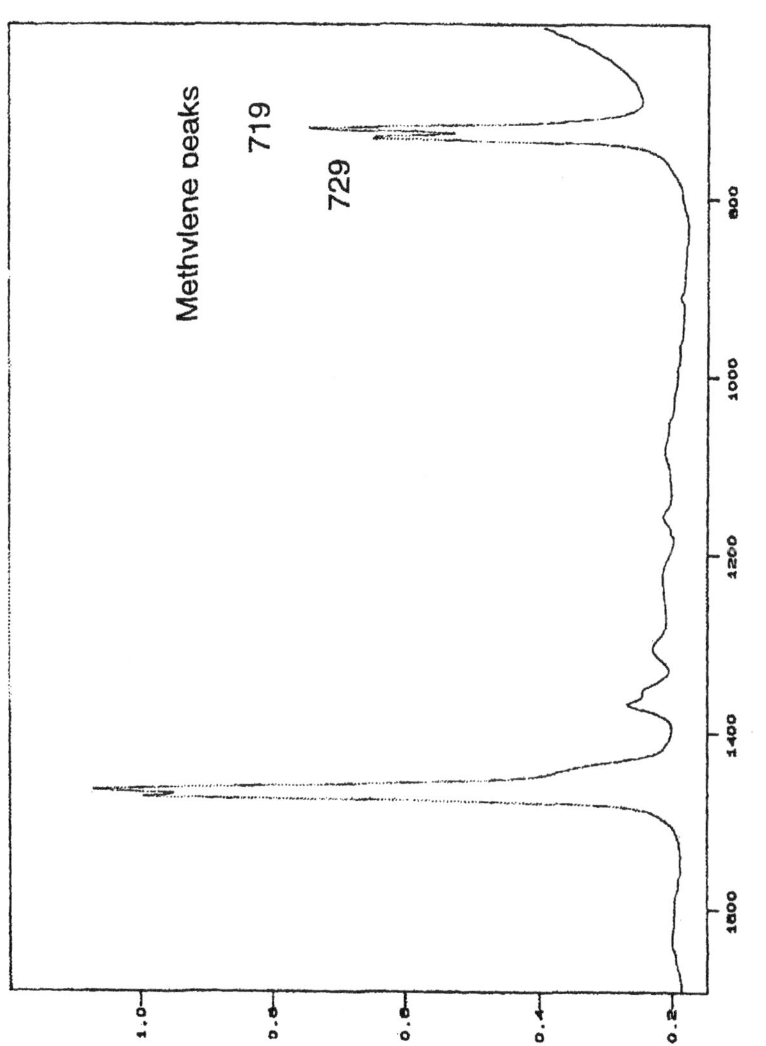

Figure 4. IR spectrum of polyethylene sample.

Summary

We have discussed the detailed synthesis and characterization of Cu(II) bis-benzimidazole complexes. The catalysts reported here represent the first homogeneous metal complexes based on copper to homopolymerize ethylene. The polymer characterization data clearly suggest that the polyethylene microstructure is highly linear with low polydispersity. The ethylene homopolymerization data is consistent with a single site polymerization catalyst.

Literature Cited

1. Ittel, S. D.; Johnson, L. K.; Brookhart, M. *Chem. Rev.* **2000**, 100, 1169-1203.
2. Britovsek, G. J. P.; Gibson, V. C.; Wass, D. F. *Angew. Chem. Int. Ed.* **1999**, 38, 428-447.
3. Boffa, L. S.; Novak, B. M. *Chem. Rev.* **2000**, 100, 1479-1493.
4. Yanjarappa, M. J.; Sivaram, S. *Prog. Polym. Sci.* **2002**, 27, 1347.
5. Mecking, S. *Angew. Chem. Int. Ed.* **2001**, 40, 534.
6. Matusi, S.; Tohi, Y.; Mitani, M.; Saito, J.; Makio, H.; Tanaka, H.; Nitabaru, M.; Nakano, T.; Fugita, T. *Chem. Lett.* **1999**, 1065. (b) Matusi, S.; Mitani, M.; Saito, J.; Tohi, Y.; Makio, H.; Tanaka, H.; Fugita, T. *Chem. Lett.* **1999**, 1263.
7. Britovsek, G. J. P.; Gibson, V. C.; Kimberley, B. S.; Maddox, P. J.; McTavish, S. J.; Solan, G. A.; White, A. J. P.; Williams, D. J. *Chem. Commun.* **1998**, 849-850.
8. Britovsek, G. J. P.; Bruce, M. I.; Gibson, V. C.; Kimberley, B. S.; Maddox, P. J.; Mastroianni, S.; McTavish, S. J.; Redshaw, C.; Solan, G. A.; Strömberg, S.; White, A. J. P.; Williams, D. J. *J. Am. Chem. Soc.* **1999**, 121, 8728-8740.
9. Small, B. L.; Brookhart, M.; Bennett, A. M. A. *J. Am. Chem. Soc.* **1998**, 120, 4044050.
10. Stibrany, R. T.; Patil, A. O.; Zushma. *Polym. Mater. Sci. Eng.* **2002**, 86, 323.
11. Stibrany, R. T.; Schulz, D. N.; Kacker, S.; Patil, A. O.; Baugh, L. S.; Rucker, S. P.; Zushma, S.; Berluche, E.; Sissano, J. A. *Polym. Mater. Sci. Eng.* **2002**, 86, 325.
12. Stibrany, R. T.; Schulz, D. N.; Kacker, S.; Patil, A. O. U. S. Patent 6,037,297 (2000); U. S. Patent 6,417,303 (2002), Stibrany, R. T., U. S. Patent 6,180,788 (2001)
13. Vyas, P. C.; Oza, C. K.; Goyal, A. K. *Chem. and Ind.* **1980**, 287-288.

14 Selected bond lengths (Å) and angles (°): Cu-N1 1.979(5), Cu-N3 1.979(4), Cu-Cl1 2.2218(19), Cl2 2.2292(19), N(1)-Cu-N(3) 141.78(17), N(1)-Cu-Cl(1) 97.58(15), N(3)-Cu-Cl(1) 94.53(13), N(1)-Cu-Cl(2) 96.32(14), N(3)-Cu-Cl(2) 97.58(14), Cl(1)-Cu-Cl(2) 139.52(7). Crystal data for **13**: $C_{36}H_{30}N_4Cl_4Cu$, M = 724.02, monoclinic, space group $P2_1/a$, a = 16.250(4), b = 15.629(8), c = 14.067(5) Å, β = 109.01(3), Z = 4, V, $Å^3$ = 3378(2), crystal size, mm 0.46 x 0.30 x 0.20.

Chapter 15

Copper-Based Olefin Polymerization Catalysts: High-Pressure ^{19}F NMR Catalyst Probe

Robert T. Stibrany

Corporate Strategic Research, ExxonMobil Research and Engineering Company, 1545 Route 22 East, Annandale, NJ 08801

Reports of nickel and palladium bisimine catalysts and, more recently, iron and cobalt tridentate catalysts has spurred the research in late transition metal catalyzed polymerizations. The polymerization of ethylene with nickel and palladium imine catalysts has generally yielded substantial branching in the polyethylenes produced. The more recent reports indicated that cobalt and iron catalysts produced far fewer branches in the polyethylenes yielded. We report the polymerization of ethylene using copper bisbenzimidazole catalysts that produce very highly linear polyethylene with the ability to incorporate certain α-olefins. High Pressure ^{19}F NMR studies suggest that these copper based systems act as single-site polymerization catalysts for ethylene and α-olefins.

Late Transition Metal Polymerization Catalysts

There has been considerable recent activity in the area of late transition metal polymerization catalysis. *(1, 2)* In the last two decades, metallocenes have revolutionized the commercial polymerization of olefins. In many cases, these catalysts are now used in place of Ziegler-Natta catalysts to produce better performing HDPE and LLDPE. Metallocenes are based on early transition metals, predominantly Ti, Zr, and Hf. These metals are very oxophilic and are very easily poisoned by polar monomers and contaminants. Catalysts with the ability to incorporate polar monomers may give rise to new high performance polymeric materials with improved performance characteristics such as high adhesion, toughness, or dyability. These catalysts may also be more tolerant of minor contaminants in ethylene polymerization. In the search for these type of polymerization catalysts, copper based catalysts have been found that produce very highly linear polyethylene while being able to incorporate certain α-olefins. *(3, 4, 5)*

Reports of nickel and palladium bisimine catalysts by UNC and DuPont, *(6, 7)* have spurred the research in late transition metal catalyzed polymerization. More recently iron and cobalt catalysts containing tridentate nitrogen ligands have also been found to be active polymerization catalysts. *(8, 9)* The polymerization of ethylene with nickel and palladium bisimine catalysts has generally yielded substantial branching in the polyethylenes produced. The more recent reports of cobalt and iron catalysts produced far fewer branches in the polyethylenes, however, early work also showed considerable chain transfer to aluminum in these systems.

We report here the polymerization of ethylene based on copper bisbenzimidazole catalysts that produce very highly linear polyethylenes of substantial molecular weight and low polydispersity. A key aspect of this work, and a particular challenge to chemists who study single site polymerization catalysts, is the characterization of the active metal site. For example, many direct and indirect probes have been used to study metal sites before, during, and after polymerization. However, catalyst species that contain ferromagnetic or paramagnetic metals offer a special challenge in that they render conventional ^{1}H and ^{13}C NMR techniques ineffective.

In the present study, high pressure ^{19}F NMR has been found to be a useful indirect method for studying these copper catalyzed polymerizations. The synthesis and structure of the bisbenzimidazole copper ^{19}F probe catalyst is described and catalyst structure-reactivity relationships with ethylene and α-olefins are discussed.

Experimental

The probe copper complex is prepared in three synthetic steps. These synthetic steps are illustrated in (Scheme I). The first step is the condensation reaction, adapted from a literature procedure, of malonic acid in polyphosphoric acid with two equivalents of 4-fluoro-1,2-phenylenediamine. *(10)* Condensation of malonic acid yields the methylene bridged bisbenzimidazole, **3**. The next synthetic step is alkylation of the bisbenzimidazole. *(11, 12)* Compound **3** can be alkylated sequentially up to four times, once each at the protonated nitrogens and twice at the methylene bridgehead carbon. In this study **3** was alkylated three times with *n*-butyliodode times to yield **4**. Finally, the alkylated bisbenzimidazole is metalated using $CuCl_2 \cdot 2H_2O$ in a mixture of ethanol and triethylorthoformate to yield **5**. Triethylorthoformate is added to effect the dehydration of the reaction mixture and aid in the crystallization of the metal complex. The solid yellow-green crystalline complex is then isolated by filtration. Suitable X-ray quality crystals were obtained by the slow evaporation of a filtered methanol solution of **5** (Figure 1).

High Pressure ^{19}F NMR Polymerization Conditions

For high pressure ^{19}F NMR studies, a custom made high pressure sapphire tube *(13)* was used to conduct probe catalyst studies. Purified toluene-d_8 was used as the solvent for the experiments. The spectra were obtained unlocked and without spinning and ^{19}F NMR were referenced to an external C_6F_6 standard. Postulated species for the polymerization using pre-catalyst **5** are presented in Table I. For the actual polymerization experiment, the NMR tube was charged with toluene-d_8, **5**, MAO, and 500 (psig) of ethylene. After mixing, the reaction was monitored at ambient temperature for 1/2 h. The reaction was then heated to 80 °C and monitored for 5 h during which polyethylene was produced. The reaction was then returned to ambient temperature and monitored once again.

Comparative Ethylene/α-Olefin Polymerization Conditions

A typical olefin polymer was obtained by an equimolar charge of comonomer followed by pressurization with ethylene to 700 psig; a non-fluorinated copper catalyst (Figure 2) solution in toluene with methylaluminoxane; Al:Cu $\Rightarrow$ 220:1, 80 °C. The resulting polymer compositional data are given in Table II.

Condensation

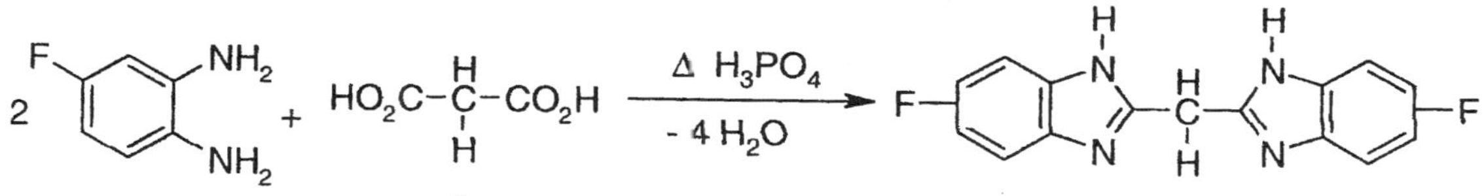

$$\Delta\ H_3PO_4$$
$$-\ 4\ H_2O$$

Alkylation

i) ex NaH ii) 3 n-butl
DMSO

1

2

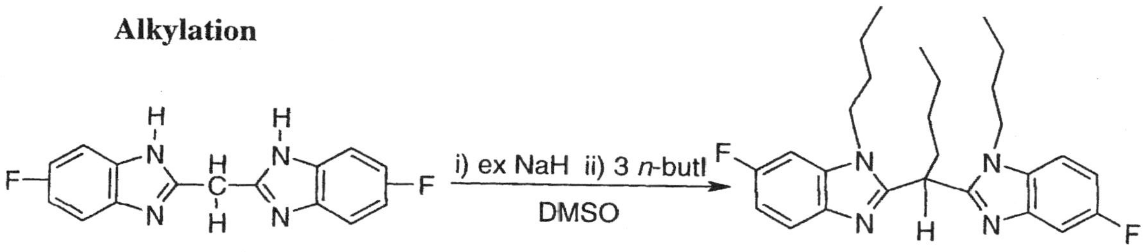

3

4

Metalation

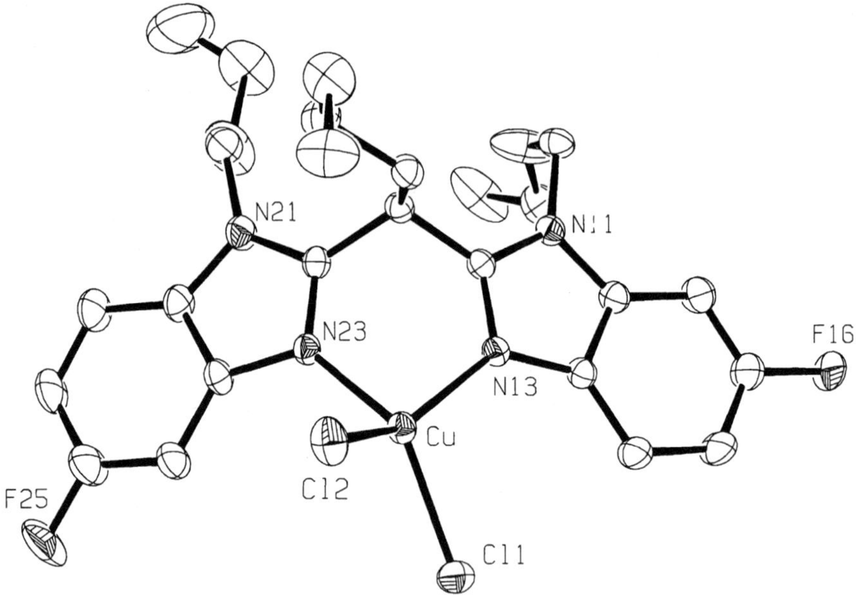

Scheme I. Synthetic steps for the preparation of the bisbenzimdazole copper ^{19}F probe catalyst.

Figure 1. Structure of 5, [1-(1-butyl-5-fluorobenzimidazol-2yl)-1'-(1-butyl-6-fluorobenzimidazol-2yl)pentane] Copper(II) dichloride. The hydrogen atoms have been omitted for clarity.

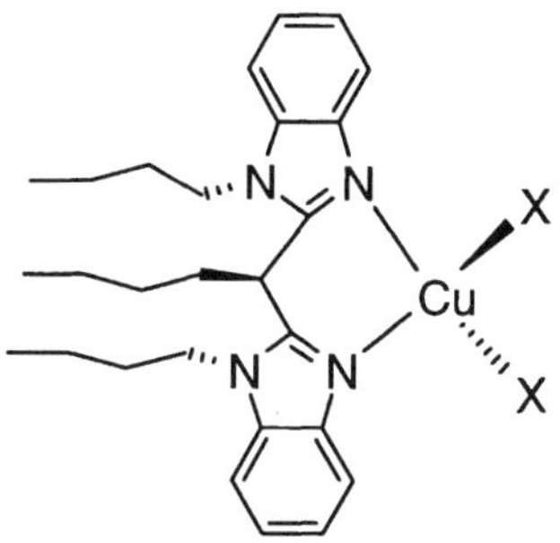

Figure 2. Non-fluorinated copper catalyst used for comparative α-Olefin polymerization study.

Table I. ^{19}F NMR Data for High Pressure Ethylene Polymerization

Exp.	Chemical shift (ppm)	Δ (Hz)	Conditions	Postulated Species
1	-118.2, -120.4	880	Ambient Soln.	Free Ligand **4**
2	-117.6, -119.9	920	Ambient Soln.	MAO/TMA Interaction **4a**[a]
3	-115.1, -119.0	1560	Ambient Soln.	Cu Complex **5**
4	-109.7, -110.9[b]	480	Ambient Temp.	Tight Ion Pair **6**
4	-112.7, -114.7[c]	800	MAO, 500 psig	Transient Species **6a**
5	-109.7, -110.9[d]	480	80 °C, 5 h	Tight Ion Pair **6**
5	-110.2, -111.2[e]	400	Polymerization	Active Catalyst **7**
6	-109.7, -110.9	480	Return to Ambient	Tight Ion Pair **6**

[a]interaction of the free ligand **4** with excess MAO/TMA in solution

[b]sharp signal

[c]minor broad short lived species

[d]sharp signal 60 % by integration

[e]broad signal 40 % by integration

Table II. Compositional Data for Ethylene/α-Olefin Copper Catalyzed Polymerization

Comonomer	Yield (g)	Mol %[a]	Mw x 10^{-3}	MWD	Tm °C
none	6.74	-	150	2.50	136
propylene	0	-	-	-	-
isobutylene	0.15	-	-	-	-
3-methylbutene	0.30	-	-	-	134
1-butene	15.20	2.2	88	3.00	122
1-hexene	14.98	0.5	511	2.19	122
1-octene	14.75	1.8	152	2.64	115
1-octadecene	13.04	1.0	309	2.63	17[b]/115
norbornene	1.51	2.7	467	2.43	120

[a] mole percent incorporation of α-olefin

[b] side chain crystallinity

Results and Discussion

The facile synthesis of the copper probe molecule 5, which is presented in Scheme I, has an unexpected synthetic result. In the alkylation step, the possibility exists for generating three isomers with respect tofluorine substitution, i.e., the 5,5', 6,6', and 5,6'. Based on the alkylation of mono-benzimidazoles, all three isomers were expected. (14) Under our experimental conditions, however, only the 5,6' isomer was observed. In Table III and Figure 3 the downfield ^{13}C NMR of 4 can clearly be assigned. The identity and purity are further supported by TLC and the crystal structure of 5. Further study is needed to determine the mechanism of the reaction

The use of ^{19}F NMR to study ferromagnetic and paramagnetic species has been slowly evolving. (15) Very few examples of this technique have been reported for studying Cu(II) paramagnetic species. (16) Since the fluorine substituents are far removed from the metal center, line broadening and chemical shift effects are minimized. The 5,6' substitution allows pairs of signals to be easily recognized. This has the added benefit of chemical shift differences between the pairs which can be greatly influenced by the metal center. Furthermore, the electron withdrawing effect of the fluorines leads to a dramatically reduction of the productivity of the catalyst, thereby allowing the observation of reaction intermediates. The use of high pressure (500 psig of ethylene) allows for the study of such intermediates under realistic polymerization conditions.

Table III. Downfield ^{13}C NMR Chemical Shifts and Assignment for 4, 1-(1-butyl-5-fluorobenzimidazol-2yl)-1'-(1-butyl-6-fluorobenzimidazo-2yl)pentane

C atom number	Chemical shift (ppm)	J (Hz)
2	152.0	-
3	137.6	-
4	119.2	10.0
5	109.4	25.1
6	159.6	34.8
7	95.5	27.4
1	134.6	12.9
2'	151.3	-
3'	141.6	12.6
4'	104.2	24.0
5'	157.2	32.4
6'	110.0	26.1
7'	109.1	10.5
1'	131.0	-
8	44.7	-

In Table I, six distinct species can be identified which would account for most of the species that could possibly be envisioned for the polymerization using the probe molecule **5**. The first three species were identified in solution without ethylene present. A distinct pair of signals was identified for the free ligand **4**. A distinct pair of signals, **4a**, was identified for the interaction of **4** with methylaluminoxane and trimethylaluminum if the copper metal were displaced by aluminum from the complex. The paramagnetic Cu(II) species, **5**, gives rise to a pair of signals with the largest chemical shift difference. A difference of 1560 Hz is significantly larger than the other 5 identified species. This is due to the strong chemical shift effect of the paramagnetic metal center. During the polymerization, only diamagnetic copper species are present. NEXAFS and epr data suggest that only Cu(I) and Cu(III) are present during the polymerization. (*17*)

In the initial polymerization activation of **5**, a short lived transient species, **6a**, is observed. This is likely one of the chloride containing intermediates that is lost during activation with MAO. After the loss of this species only one persistent pair of signals is observed, **6**. This is designated as a "tight ion pair". As previously mentioned the electron withdrawing groups dramatically lowers the productivity of the copper catalyst. In fact at ambient temperature no polyethylene is formed. The electron withdrawing effect of the fluorines causes

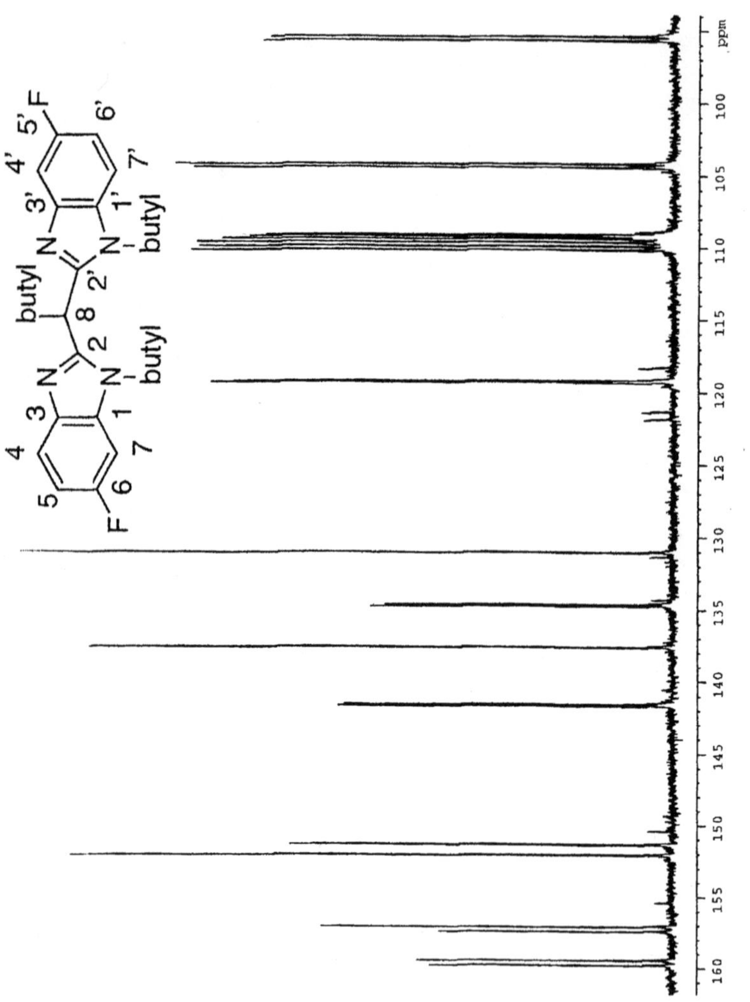

Figure 3. Downfield ¹³C NMR and atom numbering for 4, 1-(1-butyl-5-fluorobenzimidazol-2yl)-1'-(1-butyl-6-fluorobenzimidazo-2yl)pentane

greater coulombic attraction between the metal center and the anion to form a tight ion pair, which prevents an open site for ethylene binding. (*18*) Upon heating the polymerization mixture to 80 °C, during which polyethylene is formed, a second set of broadened signals, **7,** is observed. The formation of polyethylene and the expected line broadening are consistent with this set of signals being attributed to the active catalyst species. As the polymerization is returned to ambient temperature, only the pair of signals, **6,** attributed to the tight ion pair is observed. The polymerization also ceases upon cooling.

In Table II, copolymerization data are presented for ethylene and selected hydrocarbon olefins with the non-fluorinated precatalyst in (Figure 2) under identical conditions. The homopolymerization of ethylene yields a remarkable polyethylene. Not only are substantial molecular weights achieved with low polydispersity, branching is less than 1 for every 1000 carbon atoms and crystallinities of greater than 65 percent are achieved. (*17*) The addition of C_4-C_{18} α-olefins comonomers has the effect of increasing catalyst productivity by more than two-fold. This type of "comonomer effect" has been well established for a number of metallocenes. (*19*)

Another striking result of the study of this catalyst is that the addition of propylene completely inhibits polymerization. There are two likely possibilities to account for this observation. Either the propylene is forming an allylic interaction or a strong γ-agostic interaction. Further mechanistic study should reveal the nature of this interaction.

Conclusions

Fluorine-19 NMR has been shown to be a useful tool for the characterization of fluorine labeled organocopper complexes. The identification of various fluorine containing species provides a method to study copper catalyzed polymerizations. Trends in chemical shifts, splitting, and coupling all provide information about the coordination sphere of the metal center. Use of this technique to characterize homogeneous copper catalysts indicates that these catalysts are likely single site due to the emergence of specific resonances during polymerization. The simplicity of the spectra argues against a multi-site catalyst. This is consistent with the physical properties of the polymers produced.

Acknowledgments

The author thanks Ronald R. Gattie for numerous polymerizations, Raymond A. Cook for assistance with the high pressure NMR studies, Richard W. Flynn for DSC measurements, Joseph A. Olkusz for GPC analysis, and

Steven P. Rucker for polymer compositional NMR analysis. The author also acknowledges ExxonMobil Corporate Strategic Research.

References

1. Britovsek, G. J. P.; Gibson, V. C.; Wass, D. F. *Angew. Chem. Int. Ed.* **1999**, *38*, 428-447.
2. Ittel, S. D.; Johnson, L. K.; Brookhart, M. *Chem. Rev.* **2000**, *100*, 1169-1203.
3. Stibrany, R. T.; Schulz, D. N.; Kacker, S.; Patil, A. O., WO Patent 9930822, 1999.
4. Stibrany, R. T.; Schulz, D. N.; Kacker, S.; Patil, A. O., US Patent 6037297, 2000.
5. Stibrany, R. T., US Patent 6180788, 2001.
6. Johnson, L. K.; Killian, C. M.; Arthur, S. D.; Feldman, J.; McCord, E. F.; McLain, S. J.; Kreutzer, K. A.; Bennett, M. A.; Coughlin, E. B.; Ittel, S. D.; Parthasarathy, A.; Temple, D. J.; Brookhart, M. S. WO 9623010, 1996.
7. Brookhart, M. S.; Johnson, L. K.; Killian, C. M.; Arthur, S. D.; Feldman, J.; McCord, E. F.; McLain, S. J.; Kreutzer, K. A.; Bennett, M. A.; Coughlin, E. B.; Ittel, S. D.; Parthasarathy, A.; Wang, L.; Yang, Z. US Patent 5866663, 1999.
8. Small, B. L.; Brookhart, M.; Bennet, A. M. A. *J. Am. Chem. Soc.* **1998**, *120*, 4049-4050.
9. Britovsek, G. J. P.; Bruce, M.; Gibson, V. C.; Kimberley, B. S.; Maddox, P. J.; Mastroianni, S.; McTavish, S. J.; Redshaw, C.; Solan, G. A.; Strömberg, S.; White, A. J. P.; Williams, D. J. *J. Am. Chem. Soc.* **1999**, *121*, 8728-8740.
10. Vyas, P. C.; Oza, C. K.; Goyal, A. K. *Chem. and Ind.* **1980**, 287-288.
11. Stibrany, R. T.; Schugar, H. J.; Potenza, J. A. *Acta Crystallogr.* **2002**, *E58*, o1142-o1144. This method was adapted from 12.
12. Bloomfield, J. J. *J. Org. Chem.* **1961**, *26*, 4112-4115.
13. Stibrany, R. T.; Gorun, S. M. *J. Organomet. Chem.* **1999**, *579*, 217-221.
14. Mathias, L. J.; Burkett, D. *Tet. Lett.* **1979**, *49*, 4709-4712.
15. Birnbaum, E. R.; Hodge, J. A.; Grinstaff, M. W.; Schaefer, W. P.; Henling, L.; Labinger, J. A.; Bercaw, J. E.; Gray, H. B. *Inorg. Chem.* **1995**, *34*, 3625-3632.

16. Belle, C.; Beguin, C.; Gautier-Luneau, I.; Hamman, S.; Philouze, C.; Pierre, J. L.; Thomas, F.; Torelli, S.; Saint-Aman, E.; Bonin, M. *Inorg. Chem.* **2002**, *41*, 479-491.
17. Stibrany, R. T.; Patil, A. O.; Zushma, S. **2002**, ACS Meeting, Orlando, FL.
18. Richardson, D. E. In *Electron Transfer Reactions*; Isied, S. S., Eds.; American Chemical Society: Washington, D.C.; 1997: pp79-90.
19. Seppala, J. V.; Koivumaki, J.; Liu, X. J. *Poly. Sci., Part A: Poly. Chem.* **1993**, *31*, 3447-3452.

Chapter 16

Cu Catalysts for Homo- and Copolymerization of Olefins and Acrylates

R. T. Stibrany, D. N. Schulz, S. Kacker, A. O. Patil, L. S. Baugh, S. P. Rucker, S. Zushma, E. Berluche, and J. A. Sissano

Corporate Strategic Research, ExxonMobil Research and Engineering Company, 1545 Route 22 East, Annandale, NJ 08801

Cu bis-benzimidazole pre-catalysts (activated by MAO) not only homopolymerize both ethylene and various acrylates but also copolymerize these two monomer classes. Homopolymerization of acrylates and methacrylates catalyzed by CuBBIM/MAO system proceeds under mild conditions. The tacticities for acrylate homopolymers using CuBBIM/MAO do not match those expected for free radical products. Ethylene and t-butyl acrylate have been copolymerized at 600 - 800 psig and 80 °C using Cu BBIM / MAO catalysts. The products have been shown to be true copolymers by ^{13}C NMR and GPC/UV. In these copolymers, high levels of acrylate incorporation with low level of branching are possible.

The copolymerization of olefins with acrylate monomers has been an elusive goal for polymerization chemists. Acrylate monomers tend to poison Ziegler-Natta and metallocene catalysts (*1*). Late transition metal catalysts (*2-6*), in general, are less sensitive to polar species. However, the known late metal catalysts still have difficulty preparing acrylate homopolymers and/or in-chain olefin/acrylate copolymers. For example, the Grubbs neutral nickel catalyst (*7-9*) only copolymerizes ethylene with functional monomers, wherein a spacer group separates the functionality and the reactive double bond. Brookhart's early diimine Pd catalysts (*10-12*) are limited to low acrylate copolymers, which have the acrylate group at the ends of branches rather than in the chain. Brookhart's (*13*) newer diimine catalysts make in-chain polymers but they tend to be low molecular weight materials. Drent's (*14*) recent palladium catalysts also tend to produce low molecular weight, low acrylate copolymers.

We report (*15-17*) that Cu bis-benzimidazole pre-catalysts (activated by MAO) not only homopolymerize both ethylene and various acrylates but also copolymerize these two monomer classes. Unlike Brookhart catalyzed ethylene/acrylate copolymers, the Cu catalyzed copolymers are in-chain copolymers with high levels of acrylate incorporation. Low to high molecular weight products have been prepared. This paper details the Cu catalyzed homopolymerization of acrylates and copolymerizations of ethylene and alkyl (meth)acrylates and the characterization of the products formed from these systems.

Experimental

General Conditions. Toluene was purified by passing through columns of activated alumina and Q-5 copper catalyst. *t*-Butyl acrylate, *n*-butyl acrylate, methyl methacrylate, toluene, methanol and THF were obtained from Aldrich. Monomers were purified by drying over calcium chloride followed by distillation and then stored under argon in the dark over an inhibitor like phenothiazine. Mehylaluminoxane was obtained as a 30 weight percent solution in toluene from Albamarle and was stored in a refrigerator in argon glove box. Ethylene was obtained from Matheson. ^{1}H and ^{13}C NMR spectra were obtained on a Varian Unity Plus 500-MHz FT-NMR spectrometer. The solvent used were $CDCl_3$ with $Cr(acac)_3$ as a relaxation agent. Infrared spectra were measured using a Mattson Galaxy Series 5000 spectrophotometer. Gel permeation chromatography (GPC) was performed on a Waters Associates chromatograph equipped with three Polymer Laboratory mixed-bed type D columns and Peak Pro. Software. Dual detectors, namely DRI and UV-Vis @ 215 nm were used to determine the distribution of the acrylate monomer over the entire copolymer distribution. A closely matched curve from both detectors indicated a uniform distribution of acrylate and hence a true copolymer. Thermal characterization was done using a Seiko SII DSC instrument with a heating rate of 10 °C/min. The reported melting and glass transition temperatures were taken from the second heat scan after cooling to -110 °C.

Preparation of Polyethylene

A typical procedure is reported in a separate chaper (*18*).

Preparation of Poly(*t*-butyl acrylate)

A 20.1 mg (0.044 mmol) quantity of Cu(MeBBIOMe)Cl$_2$ (Catalyst 1) was added to a 100 mL round-bottomed flask in an Ar glovebox. A 10 mL quantity of toluene was added to the flask, followed by 0.11 g of 30 wt.% MAO (0.57 mmol) resulting in an yellow slurry. 7.45 g of *t*-butyl acrylate (freshly distilled from CaCl$_2$ and stabilized with 300 ppm of phenothiazine) was added to the slurry. The flask was covered with aluminum foil and the mixture was allowed to stir at room temperature for 18 hours in the dark. At the end of this time period, the reaction was quenched with 5 mL of methanol and then the polymer was precipitated out in 150 mL of acidic methanol (10%). The polymer was isolated by filtration and dried under vacuum at 40 °C, for a day. Yield: 57%. M$_n$ = 470,000, M$_w$ = 850,000; MWD = 1.8. ^{13}C NMR (ppm, CDCl$_3$): 28.2 (s, -CH$_2$-CH(COOC($\underline{C}$H$_3$) $_3$)-, 34.3-37.6 (m, -$\underline{C}$H$_2$-CH(COOC(CH$_3$) $_3$)-, 42-43.5 (m, -CH$_2$-$\underline{C}$H(COOC(CH$_3$)$_3$)-, 80.5 (m, -CH$_2$-CH(COO$\underline{C}$(CH$_3$) $_3$)-, 173.2-174.1 (m, -CH$_2$-CH($\underline{C}$OOC(CH$_3$) $_3$)-, 38% rr, 46% mr, 16% mm (by integration of methine peak).

Preparation of Ethylene/*t*-Butyl Acrylate Copolymer

A Parr reactor was loaded with 33.5 mg (0.0679 mmol) of Cu(tet EtBBIM)Cl$_2$ (Catalysts 2) followed by 35 mL of toluene, then by 2.0 mL of 30% MAO (0.01 moles) in an argon dry box to give a yellowish suspension. The 6.0 mL (5.37 g) (54 mmol) of *t*-butyl acrylate was added to give a yellow-green suspension. The Parr was sealed and set up in a hood and pressurized with 750 psig of ethylene and polymerized at 90 °C for 24 hours. The reaction mixture was cooled and quenched with MeOH. Subsequently, the contents of the reaction were added to ca 150 mL of MeOH giving a white precipitate. A 10 mg quantity of BHT and 25 mL of HCl was added, and the mixture was allowed to soak to dissolve catalyst residues. The polymer was extracted from the water phase with CH$_2$Cl$_2$ and Et$_2$O. The solvents were removed by vacuum and the polymer was dried in a vacuum oven at 55 °C to give 2.96 g of pale-green solid. ^{1}H NMR (CDCl$_3$) δ 0.7-0.85 (m, CH$_3$ end groups), δ 1.1-1.25 (m, -CH$_2$-), δ 1.4 (s, O-C(CH$_3$)$_3$), δ 2.05-2.25 (broad m, -CH). Integration of the monomer units indicates a copolymer composition of ca 67% *t*-butyl acrylate and 33% ethylene units. GPC (THF, polystyrene calibration, with DRI and UV detection at 215 nm) of a sample purified through a neutral alumina column to remove MAO and unreacted monomer: M$_n$ = 26,200, M$_w$ = 34,200. DSC showed a Tg of +4 °C and no Tm.

NH2
HO
y
OH
NH2
O
O
H3PO4
- H2O
H
N
y
N

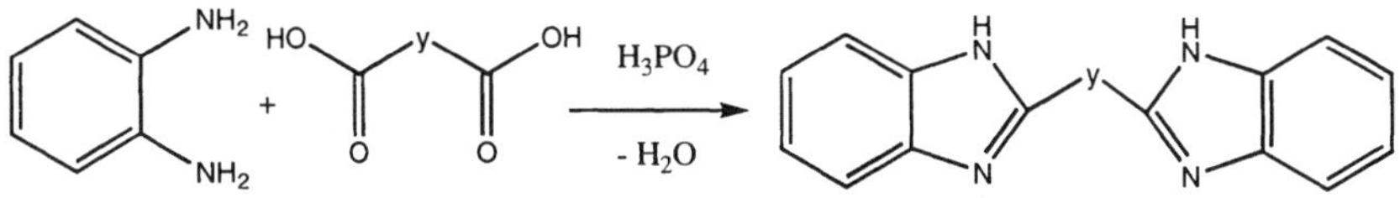

y = CHOH, CH_2, 2,2'-Biphenyl

Ia. y = CHOH
Ib. y = CH_2
Ic. y = 2,2'-Biphenyl

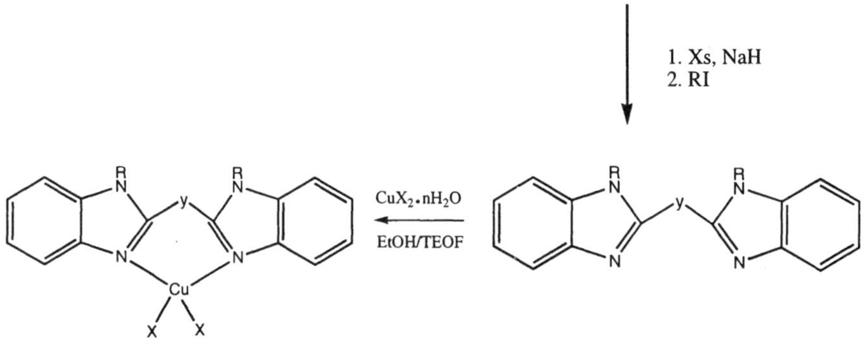

1. Xs, NaH
2. RI
R
N
y
N
Cu
X
CuX2•nH2O
EtOH/TEOF

Catalysts: 1, 2, 3a, 3b, 4Cl, 4Br

1. y = $CHOCH_3$, R = CH_3, X = Cl
2. y = $CHCH_2CH_3$, R = CH_2CH_3, X = Cl
3a. y = Biphenyl, R = CH_2CH_3, X = Cl
3b. y = Biphenyl, R = Octyl, X = Cl
4Cl y = n-Bu, R = n-Bu, X = Cl
4Br y = $CH(CH_2)_3CH_3$, R = - $(CH_2)_3CH_3$, X = -Br_2

IIa. y = $CHOCH_3$, R = CH_3
IIb. y = $CHCH_2CH_3$, R = Et
IIc. y = Biphenyl, R = Et, Octyl
IId. y = n-Bu, R = n-Bu

Results and Discussion

Cu bis-benzimidazole precatalysts (1,2,3a,3b,$4Cl_2$, $4Br_2$) are synthesized as shown in preparative scheme 1.

Activation of the Cu precatalysts (**1-4**) with MAO results in active catalysts that homopolymerize ethylene and various acrylates. Figure 1 compares the product with of various Cu catalysts for homopolymerization of ethylene and homopolymerization of various acrylates. It can be seen that the order of reactivity is t-BuA > n-BuA > MMA > E. Also it should be noted that Cu BBIM / MAO does not homopolymerize conventional free radically polymerizable monomers, such as styrene, vinyl acetate or butadiene.

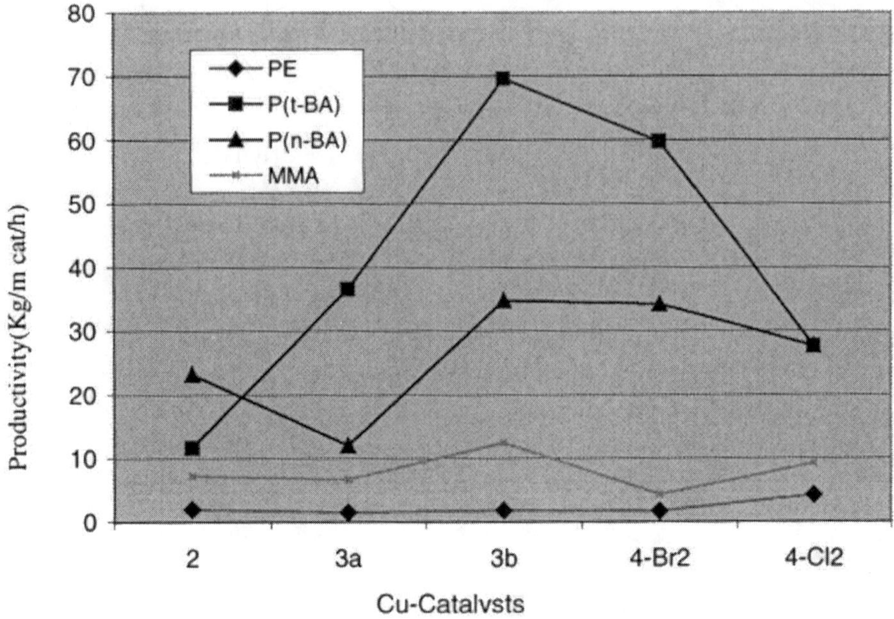

Figure 1. Experimental Conditions: Acrylates: Solvent: Toluene 15 ml, Time: 3h, Temp: 60°C Al/Cu = 250; for Polyethylene: Pressure: 750 psig, Solvent: Toluene 120 ml, Time: 3h, Temp: 60°C Al/Cu = 1500.

Table 1 shows the homopolymerization of *t*-BuA with Cu BBIM/MAO catalyst systems. It can be seen that high MW polymers are made and the tacticity do not match that for the free radically prepared p(*t*-BuA).

Ethylene and *t*-butyl acrylate have been copolymerized at 600 - 800 psig and 80 °C using Cu BBIM / MAO catalysts. In contrast to the Brookhart di-imine Pd catalysts, high levels of acrylate incorporation are possible. Also, high molecular weight, narrow MWD copolymers can be made (Table 2).

Table 1. Cu Homopolymerization of *t*-BuA
(0.045 mol of catalyst; MAO per table, Toluene, 25 °C, 18 hrs, Dark)

Catalyst	Cu/Al/*t*-BuA Mole ratio	Reaction time, h	Polymer Yield[a] %	Mn, Mw $(X10^3)$ (MWD)	Tacticity mm, mr, rr%
1	0/1/14	18	Trace		
1	1/5/940	18	5		
1	1/12/1010	18	33	280, 620 (2.2)	16,45,39
1	1/13/1320	18	57	470, 850 (1.8)	
1	1/12/980	48	77	320,640 (2.0)	
3b	1/10/980	18	8		19, 43, 38
Radical Lit.[b]	-	-	-	-	mm = 44-46

[a]GC/^{1}H NMR showed only monomer, polymer and solvent.
[b]K. Matsuzaki, T. Uryu, T. Kanai, K. Hosonuma, T. Matsubara, H. Tachikawa, M. Yamada, S. Okuzono, *Makromol. Chem.* **1977**, 11, 178. Tacticities were measured at 0-60 °C.

The products have been shown to be true copolymers by ^{13}C NMR and GPC/UV. Figure 2 shows ^{13}C NMR spectra for the Cu BBIM/MAO catalyzed E/*t*-BuA copolymers compared with P(*t*-BuA). The copolymer sequence distribution have been studies by ^{13}C NMR analysis. The methine carbons of the acrylate are present in three distinct clusters at 46.5 (EAE), 44.2 (EAA, AAE) and 42.2 (AAA) ppm. The assignment of the triad sequence is consistent with published literature (*19*). The presence of EAE, and EAA/AAE triads in the former is a clear indication of copolymer formation. Additional support for the presence of copolymers is a uniform distribution of UV activities across the MWD in the GPC/UV trace (Figure 3). A typical copolymer with 67 mole % *t*-butyl acrylate shows a unimodal peak (M_w = 108,000, MWD = 1.8) by DRI and UV (215 nm) detectors. The superimpossibility of the two types of curves obtained by different detectors indicates a uniform distribution of the acrylate for all molecular weights.

Table 2. Copolymerization of Ethylene with *t*-BuA
(0.045 mol of catalyst; MAO per table, Temp = 80 °C, 18 hrs)

Catalyst	*t*-BA Conc M	E Psig	Cu/Al/*t*-BuA Mole ratio	Polymer Yield g	Mn, Mw (X10^3) (MWD)	Triad Dist EAE:EAA/ AAE:AAA
4Cl$_2$	1.9	720	1/230/1740	6.0	85, 39 (2.2)	6:35:58
4Cl$_2$	1.9	720	1/130/1780	5.9	89, 44 (2.0)	6:34:60
4Cl$_2$	1.2	620	1/230/950	2.4		11:35:54
2	1.2	600	1/240/970	1.7		
2	1.2	800	1/240/970	1.7		

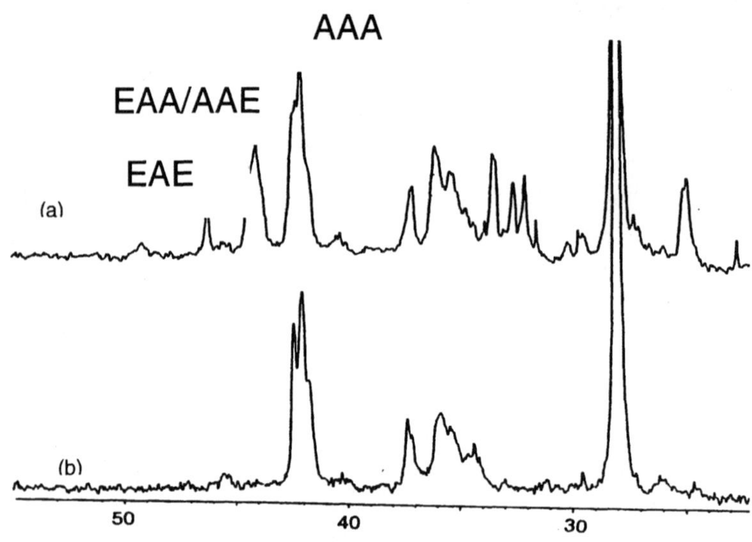

Figure 2: ^{13}C *NMR spectra of (a) ethylene/t-butyl acrylate copolymer and (b) t-butyl acrylate homopolymer*

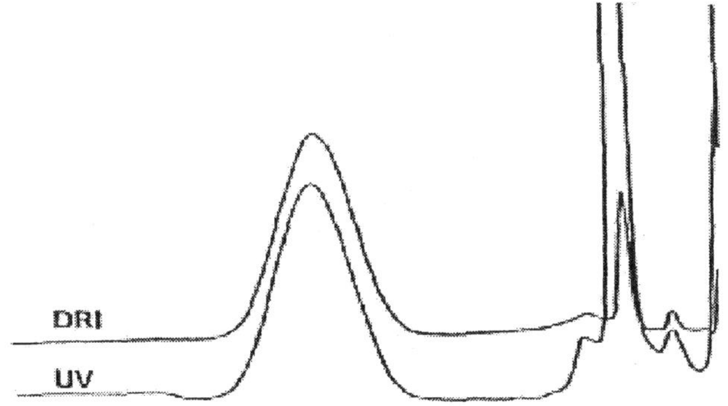

Figure 3: GPC of ethylene/t-butyl acrylate copolymer with DRI and UV (215 nm) detectors

The effect of comonomer feed on the product productivity and composition have also been studied using varying amount of *t*-butyl acrylate keeping ethylene pressure constant (700 psig). The productivity as measured by the turnover number (TON) (mol of substrate/mol of Cu) showes an unusual trend. At first, the TON decreases upon addition of low level of acrylate monomer and then increases with increasing concentration of acrylate. The copolymers obtained with very low level concentration of acrylate are essentially polyethylene, while the copolymers obtained at the higher concentration of acrylate in the feed are high in acrylate content. These results suggest a change in mechanism between the low and high acrylate feeds. This effect makes it difficult to determine the reactivity ratios for these comonomers in these copolymerizations.

In summary, Cu bis-benzimidazole pre-catalysts (activated by MAO) polymerizes various acrylates and also copolymerize ethylene with acrylates. Homopolymerization of acrylates and methacrylates catalyzed by CuBBIM/MAO system proceeds under mild conditions. The tacticities for acrylate homopolymers using CuBBIM/MAO do not match those expected for free radical products. Ethylene and *t*-butyl acrylate copolymers have been shown to be true copolymers by ^{13}C NMR and GPC/UV.

Literature Cited

1. D. N. Schulz, A. O. Patil in "Functional Polymers - Modern Synthetic Methods and Novel Structures", ACS Symposium Series 704, 1, (1998).

2. Ittel, S. D.; Johnson, L. K.; Brookhart, M. *Chem. Rev.* **2000**, 100, 1169-1203.

3. Britovsek, G. J. P.; Gibson, V. C.; Wass, D. F. *Angew. Chem. Int. Ed.* **1999**, 38, 428-447.

4. Boffa, L. S.; Novak, B. M. *Chem. Rev.* **2000**, 100, 1479-1493.

5. Yanjarappa, M. J.; Sivaram, S. *Prog. Polym. Sci.* **2002**, 27, 1347.

6. Mecking, S. *Angew. Chem. Int. Ed.* **2001**, 40, 534.

7. Wang, C.; Friedrich, S.; Younkin, T. R.; Li, R. T.; Grubbs, R. H.; Bansleben, D. A.; Day, M. W. *Organometallics* **1998**, *17*, 3149.

8. Bansleben, D. A.; Friedrich, S.; Younkin, T. R.; Grubbs, R. H.; Wang, C.; Li, R. T.; WO Patent Application 98/42665 (1998).

9. Younkin, T. R.; Connor, E. F.; Henderson, J. I.; Friedrich, S. K.; Grubbs, R. H.; Bansleben, D. A. *Science* **2000**, *287*, 5452.

10. Johnson, L. K.; Mecking, S.; Brookhart, M. *J. Am. Chem. Soc.* **1996**, *118*, 267-268.

11. Johnson, L. K.; Killian, C. M.; Authur, S. D.; Feldman, J.; McCord, E. F.; Kreutzer, K. A.; Coughlin, E. B.; Parthasarathy, A.; Brookhart, M. S. WO Patent Application 96/23010 (1996).

12. Mecking, S.; Johnson, L. K.; Wang, L.; Brookhart, M. *J. Am. Chem. Soc.* **1998**, *120*, 888.

13. Johnson, L.; Bennett, A.; Dobbs, K.; Hauptman, E.; Lonkin, A.; Ittel, S.; McCord, E.; McLain, S.; Radzewich, C.; Yin, Z.; Wang, L.; Wang, Y.; Brookhart, M. *PMSE Preprint* **2002**, 86, 319.

14. Drent, E.; van Dijk, R.; Van Ginkel, R.; Van Oort, B.; Pugh, R. I.; *Chem. Commun.* **2002**, 744.

15. Stibrany, R. T.; Schulz, D. N.; Kacker, S.; Patil, A. O., "Group 11 transition metal amine catalysts for olefin polymerization," PCT Int. Appli. WO 99/30822 (Exxon Research & Engineering Company), June 24, 1999.

16. a) Stibrany, R. T.; Schulz, D. N.; Kacker, S.; Patil, A. O. U.S. Patent 6,037,297 (March 14, 2000) and U. S. Patent 6,417,303 (July 9, 2002). b) Stibrany, R. T. U.S. Patent 6,180,788 (January 30, 2001).

17. a) Stibrany, R. T.; Patil, A. O.; Zushma. *Polym. Mater. Sci. Eng.* **2002**, 86, 323. b) Stibrany, R. T.; Schulz, D. N.; Kacker, S.; Patil, A. O.; Baugh, L. S.; Rucker, S. P.; Zushma, S.; Berluche, E.; Sissano, J. A. *Polym. Mater. Sci. Eng.* **2002**, 86, 325.

18. Stibrany, R. T.; Patil, A. O.; Zushma. See chapter entitled "Copper Based Olefin Polymerization Catalysts" in this book.

19. a) McCord, E. F.; Shaw, W. H.; Jr.; Hutchinson, R. A. *Macromolecules* **1997**, 30, 246. (b) Randall, J. C.; Ruff, C. J.; Kelchtermans, M.; Gregory, B. H. *Macromolecules* **1992**, 25, 2624. (c) Bruch, M. D.; Payne, W. G. *Macromolecules* **1986**, 19, 2712.

Indexes

Author Index

Subject Index

M